The Benefits of Imperfection

The cult of performance leads our society to emphasise the values of success and continuous optimisation in all areas. Slowness, redundancy or randomness are therefore negatively perceived. Olivier Hamant, in his book, reclaims them by his knowledge of biological processes.

What can we learn from life sciences? While some biological mechanisms certainly boast formidable efficiency, recent advances instead highlight the fundamental role of errors, incoherence or slowness in the robustness of living organisms. Should life be considered suboptimal? To what extent could suboptimality become a counter-model to the credo of performance and control in the Anthropocene?

In the face of pessimistic observations and environmental alerts, the author outlines solutions for a future that is viable and reconciled with nature.

KEY FEATURES:

- Solidly documents with a grounding in scientific facts focusing on solutions
- Explores a pragmatic way towards robustness, moving the debate beyond performance, technolatry or degrowth
- Responds to eco-anxiety by providing an engaging and viable way forward

Olivier Hamant is a researcher at the French National Research Institute for Agriculture, Food and Environment (Institut National de Recherche pour l'Agriculture, l'Alimentation et l'Environnement, INRAE) at École Normale Supérieure de Lyon. As an interdisciplinary biologist, he has published approximately 100 scientific articles, notably on the development of plants and their ability to perceive their own shape. He also heads the Michel Serres Institute and is involved in training programmes on the new relationship between humanity and nature. As an independent author, he published *La Troisième Voie du Vivant* (Odile Jacob, 2022) / *The Benefits of Imperfection* (CRC Press, 2025), *Antidote to the Cult of Performance* (Gallimard, 2023) and *De l'Incohérence* (Odile Jacob, 2024, untranslated).

The Benefits of Imperfection

Biology, Society, and Beyond

Olivier Hamant

CRC Press is an imprint of the
Taylor & Francis Group, an **informa** business

Designed cover image: Olivier Hamant

First edition published 2025
by CRC Press
2385 NW Executive Center Drive, Suite 320, Boca Raton FL 33431

and by CRC Press
4 Park Square, Milton Park, Abingdon, Oxon, OX14 4RN

CRC Press is an imprint of Taylor & Francis Group, LLC

ISBN: 978-1-032-83220-3 (hbk)
ISBN: 978-1-032-84035-2 (pbk)
ISBN: 978-1-003-51091-8 (ebk)

DOI: 10.1201/9781003510918

Typeset in Times
by Apex Covantage, LLC

Contents

Acknowledgements

A warm thank-you to all at the École Normale Supérieure de Lyon, who have been my partners on this Anthropocene journey from the very start: Ioan Negrutiu, Stéphane Grumbach, Julie Le Gall and Patrick Degeorges. The topic of suboptimality emerged out of our discussions, and within the particularly fertile environment at the Anthropocene campus of the House of World Cultures in Berlin. The idea of writing an essay on this subject took hold following friendly invitations from Beate Geissler and Oliver Sann, Nathalie Ergino (the Villeurbanne Institute of Contemporary Art), Cyrille Noirjean (the International Lacanian Association), Zigor Hernandorena (Ubisoft), Michel Lussault (the Lyon Urban School) and Béatrice et Virgile Viasnoff (the Association for the Promotion of Management in Singapore). I'm grateful to the members of my research team, to my colleagues of the Plant Reproduction and Development Laboratory and of the Michel Serres Institute, and to my collaborators, for many conversations that helped to shape these thoughts. I also want to thank Mathias Pessiglione, Laurent Cohen and Marc Kirsch for their help in making this project a reality, alongside Odile Jacob. I am grateful to the following persons for suggestions and support on this path to suboptimality: Philippe Barraud, Julien Barrier, Olivier Barrière, Dominique Bourg, Valérie Colomb, Pierre Courtois, Pablo Jensen, Jindra Kratochvil, Dennis Meadows, Florence Meyssonnier, Bruno Moulia, Raphaëlle Raynaud, Pierre Thomas and David Vallat. Lastly, I wholeheartedly thank my very first readers for their useful and pertinent feedback: Pierre Cornu, Barbara Escande, Claire, Pascal, Philippe, Michèle Hamant and Sowmya Kadandale. The original French text was translated into English by TextMaster.

While I am indebted to all of them, it goes without saying that this essay is the sole responsibility of its author.

1 Preamble, as an Executive Summary

How to coexist? Since the dawn of time, humans have formed prosperous communities through their social skills. By establishing shared and accepted rules, which restrict individual freedom, humans have ensured the viability of the group. Today, our global predation on natural resources is generating so many negative consequences that it is challenging our viability on Earth. The social contract is hitting the planetary boundaries. Therefore, the challenge of this century will be to invent a social contract that is extended to our interactions with the world, that is with the "non-humans". This is what Michel Serres calls *the natural contract*.[1]

Such a project cannot be reduced to a form of superficial bio-inspiration. It must involve a profound knowledge of the mechanisms of life on Earth. An education to live on Earth. However, the complexity, and sometimes extravagance, of life makes this exercise particularly difficult. Like many people, I am fascinated by the diversity and harmony of living beings: shapes, colours, movements, communities, etc. Within my biology research team, and with my colleagues, I have had the opportunity to analyse certain molecular and cellular mechanisms underlying plant development. Far from disrupting harmony, this exploration reveals other facets of the living world, more intimate and often universal in character, since they are shared by all species. In other words, the detailed study of living beings does not break the charm and beauty, but rather expands it. What do we learn by compiling biological research studies? What do recent discoveries on living systems teach us, from the largest to the smallest scales? Here is a synthesis in the form of three fundamental pillars.

First of all, life is essentially circular. Living beings are all part of the Earth's cycles: the water cycle, the carbon cycle, the nitrogen cycle, etc. Conversely, this highlights the anomaly of our modern societies. Indeed, humans have built a world largely based on the accumulation of goods without recycling: once their full life cycles are considered, our products are primarily waste. The advent of the circular economy is a first step towards reconciling our lifestyles with the Earth's cycles. However, it would be naive to think that circularity of human activities could be limited to eco-design or recyclable material. Circular thinking mainly implies to revisit our socio-economic system by considering the feedback of our activities, including those that seem the most virtuous in the short term. This is a much more complex exercise. Experience shows that we create many counterproductive rebound effects, sometimes with the best of intentions. A principle of extended circularity could at least help us to question, or even filter, the many solutions of so-called sustainable development. Finally, the cycles of life open an infinity of possibilities and integrate the (very) long time. To live is to regenerate.

DOI: 10.1201/9781003510918-1

A second pillar of life is its collective behaviour. Throughout evolution, group viability strategies have been favoured over individual comfort. Whether it is a tree in a forest, a fish in a shoal, or a cell in a tissue, living beings often constrain their individual performance to allow the survival of the group. This observation should also question our choices. In our globalised societies, individual competitiveness is glorified. Our modern economic, scientific or technological heroes are presented as precursors and role models. In contrast, the social component is most often presented as a "burden" or as a charity to help the marginalised fringes of society. Looking at living systems forces us again to take a different standpoint: it is rather our selfish individualism that makes us marginal on Earth. Faced with the human and environmental challenges ahead, we will most likely have to reappropriate and expand the notion of society, not only between humans but also with non-humans. Connecting with these new partners is essential to our sustainability on Earth and may even offer an engaging path forward. To live is to coexist.

The third and final pillar of life is the most important to me because it makes the first two operational. Living beings do not focus on efficiency, but on robustness. What is this about? Robustness can be defined as the ability for a system to maintain its stability (in the short term) and its viability (in the long term) despite fluctuations. In other words, living organisms are selected during evolution on their ability to live with environmental fluctuations. How does such robustness emerge? This is perhaps where the surprises are the most stimulating: the robustness of life emerges from variability, heterogeneity, slowness, delays, errors, randomness, redundancy, inconsistencies, etc. In short, the robustness of living beings is not a quality added to efficiency; robustness emerges from intrinsically and locally inefficient processes, that is *against* performance. How can this work? The apparent underperformances of living beings open up a great deal of leeway, which in turn feeds adaptability. For example, consider the ability of forests to regenerate after a fire or our own ability to survive fasting for two to three weeks. But that's not all. The interactions between these underperformances create an internal equilibrium, a form of autonomy, providing the conditions to get across the vagaries of the environment. In summary, our best shield against external fluctuation is our own internal fluctuation. This principle of robustness has become vital today. Indeed, the climate crisis, the collapse of biodiversity and the multiple physico-chemical modifications of our ecosystems herald an increasingly unpredictable world in the future. Our obsession with efficiency and control is locking us into a narrower path. The process of optimisation is weakening us. On the contrary, the goal of robustness allows us to think the unthinkable because the ability to react and to adapt exists. To live is to resist.

To recap: (i) circularity implies underperforming in the short term to avoid the pitfalls of long-term feedback; (ii) collective behaviour implies sometimes underperforming individually to ensure the survival of the group; (iii) robustness is in essence the path to sustainability, and it is built against performance. Life is therefore telling us one thing three times: stop running towards efficiency!

In a society where the injunction to perform is omnipresent, life opens another path, where progress would no longer be guided by performance increments, but by robustness. It is no more and no less than another way to live on Earth. It is a societal project to become Earthlings again. But why would this "way of life" be so relevant or so urgent? What would be its mechanisms and limits? Before entering into a form of critique of the rationality of our contemporary human societies and before discussing such a counter-model, let us first consider our epoch, the Anthropocene.

2 The Age of Performance

We are the citizens of a new epoch, *the Anthropocene*.[2,3] According to most experts, this is a geological age that began at some point between the First Industrial Revolution and the start of the "Glorious Thirty". Alternatively, it may be considered the final stage of the Holocene, which began following the last ice age, 11,700 years ago. While the very term *Anthropocene* is a subject of debate, its existence may also force us to dig deeper into the roots of our modern civilisation and its various dimensions.

2.1 A QUESTION OF SCALE

The Anthropocene describes a fundamental characteristic of contemporary civilisation as we know it: human activities now have an impact on a planetary scale. In turn, threatened nature becomes more threatening: humans are confronted by their global limits. As sociologist and philosopher Bruno Latour puts it, "the Earth is finally round".[4]

As is the case with the word *America*, attributed in honour of Amerigo Vespucci, who grasped the significance of Christopher Columbus's discovery in spite of the fact that only a small fraction of the new continent was yet known, the term *Anthropocene* emerged prior to the realisation of the magnitude of the associated implications.[2] Amounting to much more than a description of the human impact on Earth, the formalisation of the Anthropocene concept acknowledges the finite nature of the world, the boomerang effects of a deteriorated natural environment on human civilisation and, more broadly, the end of modernity.[5] The Anthropocene thus also serves as a framework for the interpretation of the many changes that are currently taking place in our civilisation.

Let us begin by turning our attention to some figures that illustrate the change of scale brought about as a result of human activity within the industrial era. The production of copper multiplied 27-fold between the years 1900 (495,000 tonnes/year) and 2000 (13.2 million tonnes/year). The production of aluminium increased by a factor of 3,600 between the years 1900 (6,800 tonnes/year) and 2000 (24.3 million tonnes/year), while the production of plastics multiplied by 7,500 between the years 1925 (20,000 tonnes/year) and 2000 (150 million tonnes/year).[6] It should be noted that throughout the course of the 20th century, human population *only* increased by a factor of 3.7.

Looking beyond the extraction of materials and production of goods, the Anthropocene is even more plainly evident when considering the landscape: 83% of non-frozen land mass shows evidence of human impact.[7] Considering this in closer detail, 41% of continental surfaces that are not covered with ice are devoted to agriculture, 50% of territories are fragmented by roadways and rail lines, and more than 60% of major rivers are interrupted by one or multiple dams.[8] These modifications naturally have manifold repercussions, with regard to not only the biosphere but

DOI: 10.1201/9781003510918-2

also the planet itself. NASA has even suggested that water reservoirs constructed by humans are so numerous that they are actually slowing the Earth's speed of rotation![8]

Far from being linear, this impact shows an increase over time, with a very marked acceleration in recent decades. As such, since 1950, the number of major dams has multiplied by a factor of 5, the production of paper by 7, the number of motor vehicles by 7, the use of fertilisers by 10 and the number of journeys made for the purpose of tourism by 40.[9,10] The chemist and Earth system analyst Will Steffen has shown that the growth of most socio-economic parameters (water use, fertiliser consumption, paper production, etc.) switched from an arithmetic behaviour to a geometric one after the Second World War. Not only is this *great acceleration* very recent, it also marks a bifurcation of the trajectory of our civilisation compared to the last millennia. In other words, around 1950, we left the regime that had prevailed since the Neolithic period. For this reason, the end of the Second World War is often considered as the beginning of the Anthropocene, with the explosion of the first atomic bomb in 1945 as a stratigraphic maker.

These figures are not necessarily set to decrease. The world's 33 leading banks have invested 1.9 billion dollars in oil companies since the signing of the Paris Agreement in 2015.[11] The use of fossil fuels is heavily subsidised, and these subsidies are actually increasing on a global scale (4,700 billion dollars in 2015, 5,200 billion dollars in 2017[12]). Before the Covid-19 crisis, ExxonMobil was planning to pump 25% more oil and gas in 2025 compared to 2017.[13] Similarly, annual vehicle sales were expected to double between 2010 and 2030.[14]

2.2 A GLOBAL ISSUE

Mirroring the acceleration of human activity, ecosystems are changing at a rate previously unseen at any point throughout the Earth's history.[15] In this context, climate change is often raised as a banner by the media. Here follows a brief overview of the situation.

The incoming radiation from the sun is absorbed by the Earth's surface, which in turn re-emits infrared radiation. The atmosphere absorbs 95% of it, thanks to a small fraction of the atmospheric gases that are capable of vibrating. This is not the case for dominant gases, oxygen (O_2) or nitrogen (N_2), which are symmetrical and do not vibrate. On the contrary, gases with at least three atoms, such as water vapour (H_2O) and carbon dioxide (CO_2), do vibrate and absorb infrared radiation. These so-called greenhouse gases represent less than 1% of the atmosphere. Although they are in the minority quantitatively, their impact is major. Indeed, once absorbed, infrared radiation is re-emitted by these gases towards the ground and this new energy charge leads to the greenhouse effect. It has been calculated that without an atmosphere, the Earth's surface temperature would be 30°C lower. Water vapour would contribute 20°C, and CO_2 would contribute the remaining 10°C.

By studying bubbles of air trapped in the polar ice cores, it has now been established that the atmospheric content of CO_2 has been fluctuating between 180 and 300 parts per million by volume (ppmv) over the past 800,000 years. In 1950, measurements indicated a concentration of 311 ppmv, while in 2024 the figure stands at 428 ppmv.[8] The link between the atmospheric content of CO_2 and temperature is now very well established. Earth's average surface temperature has already increased by about 1.3°C since 1880. The potential increase by the year 2100 may range from +1.5°C to +7°C, according to various scenarios produced by the Intergovernmental Panel on Climate Change (IPCC).

The objective of reducing CO_2 emissions, with a view to limiting the warming of the planet to just 2°C by the end of the 21st century, has seen parallels drawn with the interglacial Eemian period, that is, a relatively warm (+2°C) period spanning 15,000 years towards the end of the Pleistocene (which precedes the Holocene), some 120,000 years ago. At this time, the sea level was 6 to 9 metres higher. The presence of dislodged boulders weighing more than 1,000 tonnes is evidence of very violent storms. By way of comparison, waves from the worst storms in recent decades have moved boulders weighing a maximum of 100 tonnes.[16]

The Eemian was 2°C warmer due to cyclic changes in the Earth's orbit (also known as Milankovitch cycles): the atmospheric concentration of CO_2 was just 280 ppmv. In order to find a true equivalent to an increase of 2°C due to an accumulation of greenhouse gases, one actually has to go back to the Middle Miocene, around 15 million years ago, which had an atmospheric concentration of CO_2 close to the current concentration.[17] Jump forward 15 million years, and the change occurring within the space of about two centuries is thus extremely brutal, geologically speaking.

While the issue of atmospheric CO_2 receives a great deal of media coverage, the oceans represent another ticking time bomb. Indeed, they take longer to manifest their deteriorated condition because of their planetary buffering role. One-third of the CO_2 released as a result of human activity is dissolved into the oceans (525 billion tonnes since the beginning of the industrial era). At first, this may appear to be good news, as it is the atmospheric concentration of CO_2 that causes the temperature to rise. Unfortunately, in its dissolved form, CO_2 acidifies the surface waters of the oceans, resulting in manifold collateral damage to ecosystems. Based on the chemical analysis of fossils, the oceans are thought to be just as acidic as they were 14 million years ago.[18] However, some authors surmise that the oceans have all but reached their maximum limit in terms of their ability to sequester atmospheric CO_2. This would mean that the CO_2 emitted today would no longer be regulated by the ocean carbon reservoir, thus bringing about an acceleration in global warming.[8]

In addition to the climate crisis, the chemical pollution of our environment regularly makes big media headlines. For instance, it is estimated that 300,000 tonnes of mercury have been released into the environment as a result of human activity on Earth as a whole. The effects of this are evident in the large predators of the Arctic Ocean, with creatures presenting mercury content levels that are 12 times greater than those found in the pre-industrial era.[8] More generally, the oceans have become a global rubbish dump. While the volume of oil released in oil spills has significantly decreased in recent decades, oceans are becoming filled with plastic waste. By 2050, they could host more plastic than fish (by weight).[19, 20] Plastic pollution is not limited to the oceans; it can also be found in continental soil and water. Today, with the exception of fossil aquifers and ice caps, there is no water on Earth without plastic. It then contaminates all food chains: all animals on Earth have nanoplastics in their blood. With a more global lens, humans have produced 140,000 new synthetic molecules. By way of comparison, just 250,000 natural molecules are currently known in plants. It goes without saying that this vastly increases the possibility of unexpected interactions and consequences for the biosphere.[21]

2.3 THE COLLAPSE OF BIODIVERSITY

Global warming and pollution have obvious impacts on flora and fauna. However, non-humans are also eliminated by other human activities, including destruction of habitats, predation on natural resources and introduction of invasive species.[8]

According to the United Nations Food and Agriculture Organization (FAO), a forest area equivalent to half of the surface area of the United Kingdom disappears every year, which amounts to 13 million hectares.[22] To avoid an unbalanced perspective, it should be noted that 3 million to 5 million hectares of forest are replanted each year. This consequently replaces but only a fraction of the lost forestland. Furthermore, the new plantation only accounts for a few species and does not constitute a qualitative replacement for the primary forests, where the diversity of species is significantly superior to that found in anthropogenic woodlands. According to the World Wildlife Fund (WWF), half of the Amazonian rainforest is expected to vanish by the year 2050.[23]

One-third of insect species are endangered. 2.5% of insects (weight by weight) disappear each year, a rate that would see them vanish altogether by the year 2100.[24] With regard to terrestrial vertebrates (mammals, birds, reptiles, amphibians), 65% (weight by weight) are domestic animals (mainly pigs, chickens, cows), 32% are human beings, leaving just 3% to account for all of the remaining wild vertebrates. A plant or animal species becomes extinct every 20 minutes, which corresponds to a rate 100 to 1,000 times higher than the base level. In the case of the North American passenger pigeon, it took only half a century for hunters to wipe out 2 billion to 3 billion individual birds, leading to their ultimate extinction in 1914.[8]

Behind the mind-boggling figures, the momentum of this crisis is a matter of great concerns because this so-called *great acceleration* began after the Second World War. A period of 80 years corresponds to a fraction of a second when compared to the estimated 315,000-year history of *Homo sapiens*.[25, 26] Over the past 500 million years, there have been only five mass extinctions, which involve the disappearance of 75% of species. Within the short period of the Anthropocene, our trajectory is already headed towards a sixth mass extinction of human origin.[27, 28] More specifically, the following species are at threat of extinction today: 40% of amphibians, 34% of conifers, 33% of corals and 25% of mammals.[29] In fact, while animal species were drastically reduced during previous mass extinctions, many plant species managed to survive. In the event of the sixth mass extinction to come, we are moving so quickly that many plant species are vanishing.[30] This includes even domesticated plants, 200 of which may soon join the red list of endangered species.[31]

The collapse of biodiversity has a multitude of consequences for our civilisation, particularly due to the many "free services" provided by ecosystems. For instance, 75% of agricultural crop production is dependent on pollination. The production of drinking water requires several depollution steps realised by plants and filter-feeding animals. Soil maintenance requires the production of biomass by plants and the activity of decomposers. Ecosystems also limit environmental fluctuations, such as flooding. Furthermore, biodiversity is our best shield against pandemics. Conversely, as their presence greatly outnumbers the global terrestrial fauna in mass and because they are genetically more homogeneous, domesticated animals and humans mathematically become the main warm-blooded hosts of the viruses today.[32]

The aggregate value of these ecosystem "services" has been assessed, and it far exceeds the world gross domestic product (GDP).[33] A mass extinction would therefore come at a cost, with some of these free services having to be replaced with human activity. Is this a bad science fiction scenario? No, because this is already happening. In particular, humans have become new pollinating animals in the Anthropocene. In light of the disappearance of bees, the pollination of some fruit trees in the north of China is now performed by human workers.[34]

These figures reveal the moral, and even legal, dimension of the problem: the human responsibility for an ongoing ecocide. However, they do not, strictly speaking, constitute warnings. After all, Earth has already experienced mass extinctions in the past, humans have endured ice age, CO_2 is a nutrient that is essential to plant growth and ecosystems always manage to regenerate in the end. One could even say that this difficult period of time for the planet is the price to be paid for the economic take-off of developing countries, and for the better days to come. So why ought these figures to be of concern to us?

2.4 A WARNING

Based on the measurements outlined earlier, and through the use of computer modelling tools, it is possible to predict future trajectories. The most well-known of these models is without doubt World3.

At the end of the 1960s, the Club of Rome, a think tank that brings together scientists, economists, industry leaders and government officials from 52 countries, finds itself confronted with the upcoming depletion of non-renewable resources, following centuries of questions raised regarding our capacity to manage our natural heritage.[35] It consequently commissioned Dennis and Donella Meadows at MIT with the drafting of a prospective report on the consequences of resource scarcity on other parameters affecting our civilisation, such as demography, food supply, services and global pollution. In World3, different types of scenarios are projected: *business as usual*, and a range of more proactive solutions utilising technologies that enable a more sustainable exploitation of natural resources. In qualitative terms, all scenarios produce the same result: a socio-economic tipping point set to occur before or around 2050.[36] The Meadows report thus puts to rest the idea of infinite growth. It constitutes a first general formalisation of our global natural resources and of the required adjustment of human needs. In spite of the many criticisms of the model at the time, it must be concluded that 50 years on, its predictions are coming more or less to fruition.[37, 38]

While the Meadows report raises the issue of a limit to human development, *planetary boundaries* define the "safe operating space for humanity". In 2009, Johan Rockström (Sweden) and Will Steffen (Australia) presented a report identifying nine planetary boundaries: climate crisis, biodiversity collapse, disruption of the nitrogen and phosphorus cycles, ocean acidification, loss of arable land, freshwater availability, ozone layer depletion, atmospheric aerosol concentration and chemical pollution.

Based on existing estimations, this original report indicated that we have already crossed three of these boundaries (climate change, biodiversity loss, nitrogen and phosphorus cycles).[39, 40] A 2022 re-evaluation now states that three more boundaries have been crossed (loss of arable land, freshwater availability, chemical pollution).[41] When you leave the space in which the safe operation of humanity is viable, what happens?

Crossing planetary boundaries would quite literally mean entering *terra incognita*, and in particular would produce non-linear effects that are very difficult to predict. In order to understand the related implications, let us consider the climate crisis by way of an example.

The geological history of the Earth demonstrates the existence of stable states at +4°C or +5°C, raising fears that there exists a global warming threshold beyond which we will make an irreversible transition to a much warmer world.[42] One might, at this point, cite a number of instances of positive feedback that exacerbate initial global warming, such as the melting of ice increasing the absorption of the sun's radiation on the ground (albedo effect), or the thawing of permafrost, which results in the emission of large quantities of methane, another greenhouse gas.[43] As a result of this feedback, we are thought to be at a tipping point – at the beginning of a transition to a new and significantly warmer world. This possible irreversibility could be the most worrying component of the climate crisis: Would a 4°C to 5°C increase in temperature be too great an attractor, rendering it impossible for the temperature to easily return to the cruising temperature of the Holocene, the period spanning the last 11,000 years during which relative climate stability has enabled the emergence of agriculture and of our civilisation?[42]

Last, this global approach could suggest that the threat only concerns non-renewable resources. Renewable resources are also exhaustible, as we have seen with the collapse of biodiversity. The decline of renewable resources also affects our management of agricultural systems. In particular, the quantity of drinking water or arable land is limiting the future development of the human population.[44] In France, a range of 40,000 to 90,000 hectares of arable land were lost to infrastructure or dwellings between 2000 and 2012.[8] The production of biomass by ecosystems, for the purpose of their regeneration, has been reduced to never-before-seen levels as a result of human consumption, reaching as high as 90% in areas used for large-scale arable farming, and approaching 50% on a global level.[45] Consequently, this biomass is insufficient for the regeneration of soils and ecosystems. The formalisation of "overshoot day" brought this issue to the attention of the general public. It is equal to a territory's capacity for biological production, divided by the human ecological footprint within this territory, with the total then multiplied by 365. In 2019, the European Union reached this day on 9 May, which is to say that from the following day onwards, Europeans were living on credit: they were consuming more than their territory had to offer.[46] As a comparison, overshoot day for the countries of the modern-day European Union was 13 October in 1961.[46] To put it in other words, if all human beings were living as Europeans are, it would take 2.8 planets to provide for their needs. It is very difficult to conceive where an ecological debt such as this could lead. A famous study suggests that these imbalances herald a collapse of ecosystems in the relatively short term.[47] The currently ongoing sixth extinction of the species would be the prelude to this.

Our recent history has shown us that establishing peace is much more difficult than winning the war. In the Anthropocene, the matter is even more complicated: we will have to prepare to make peace with the planet, knowing that we are first going to lose the war. Is this word too strong? In *The Natural Contract*, Michel Serres actually goes one step further. Indeed, "war" still refers to rights. For example, "war crime", a violation of the law of war, implies the existence of a legal framework. Having signed no contract with nature, what we are doing to Earth is actually closer to ruthless and ruleless violence.

Before asking the question, "*what shall we do?*", we must ask ourselves the question as to "*why?*". Why, primarily over the course of the past two centuries, have humans set about mistreating their environment with such enthusiasm and determination?

2.5 A RACE TOWARDS OPTIMISATION

The Anthropocene is an ambivalent term. On the one hand, it reflects a form of arrogance. Humanity would have become a new geological force, continually pushing the limits of time and space to further extremes, digging mines that are increasingly deep, exploiting sea beds in waters that are increasingly more polar or more abyssal, using resources day and night as shown in the pictures of planet Earth illuminated by its cities at night. But on the other hand, we are beginning to feel the global negative feedback from our planetary predation. We are polluting the ocean to ever-increasing depths, with plastic having been recovered from the bottom of the Mariana Trench at a depth of more than 10 km.[48] We are polluting the atmosphere at increasing altitudes, with up to 37,000 pieces of space debris (each measuring more than 10 cm) left in orbit.[49, 50] The viability of our civilisation is threatened. In the end, the Anthropocene is both an epoch when we claim to control the planet and a time when we lose control. How can we explain this paradox?

It is my hypothesis that our view of *optimisation* is a fundamental driving factor of the Anthropocene. In its broadest sense, as applied here, optimisation is reduced to the improvement of performance. Performance itself can be defined as the sum of efficacy (achieving one's goal) and efficiency (making the best use of available means). When we optimise, we want to achieve our goals with the least possible means: yield optimisation in agriculture (and in all economic sectors), land optimisation in urban planning, code optimisation in information technologies, tax optimisation in the fiscal world, etc. It is therefore not a question of simple maximisation (maximum improvement of performance), but in a more subtle way, of a certain rationality that legitimises performance.

The rise of engineering reflects a world driven primarily by efficacy and efficiency gains. In the Anthropocene, we have created engineers for most technical and economic functions: mechanics, aeronautics, electricity, energy, chemistry, biotechnology, geology, mining, military, naval, nuclear, information technologies etc. Engineering has also invaded the social sciences: management, social security, pedagogy, behaviour etc. Finally, everything could be optimised.

Some will say that this definition of optimisation, although very common, is not correct. When optimising, one must include all the parameters of the system and instead find a compromise. This more accurate definition has strong theoretical foundations, for example in the Pareto optimum, which describes a society where the well-being of a given individual cannot hinder that of another. In response, I insist on the fact that we must confront the reality of a definition of optimisation reduced to a gain in efficacy and efficiency, most often within an anthropocentric framework. The vision of a human optimisation that would include all terrestrial parameters is utopian: we always optimise a process within a very restricted framework. The catalogue of engineering sub-types mentioned earlier shows that we operate essentially in silos. To consider that optimisation is inclusive by default is a denial of human reality and its limits.

Furthermore, considering optimisation for what it really is functionally in our society reveals the fundamental hiatus of our time: when we implement optimisation in the form of efficacy and efficiency gains, we claim to improve our well-being at a reduced cost. In reality, we are not moving away from the concept of human progress driven by performance. As we shall see, the 21st century could constitute a profound revolution by no longer orienting progress towards performance gains, but towards maintaining and ensuring robustness. This new "way of life" is shared among all living beings on Earth.

Another criticism I hear is that efficacy and efficiency are often associated with sobriety in order to design an environmentally responsible and sustainable path. Once again, this masks certain ambivalences. The emphasis on efficiency, for example, boasts a form of frugality, but in reality, implies lean production and reduced leeway, in a framework that remains essentially extractivist. Similarly, efficacy reinforces the idea of a controlled trajectory where well-defined objectives need to be reached. Thus, optimisation embeds a belief in control.

Is this belief well founded? The emphasis on efficacy implies that the defined objectives are necessarily the right ones. But in an increasingly unstable environment with a literally unthinkable future, how can we be so sure that this is still true? Similarly, efficiency builds on short-term increments first. It rarely tackles systemic and long-term implications. Human life expectancy confines us to a fragmented and reductionist view of the world and for durations that are very short when compared to geological time. Optimisation should rather be viewed as the illusion of control.

To me, optimisation is thus the fuel of an ambivalent Anthropocene, between displayed arrogance (human development) and repressed humility (the denial of planetary and human limits). More simply, such a take on the Anthropocene allows us to question our relationship to performance in a more universal way.

The improvement of performance has become a universal credo. From a very young age, games such as the classic board game Monopoly instil the idea of performance increments. Sports competitions are tributes to performance, with many multiplier effects, such as the mirage of high salaries, doping or money laundering through betting. An ultimate symbol of this tropism for performance is the augmented human. This is not science fiction. We have indeed already begun to "improve" on a huge scale, and this starts from school age. For instance, the number of children taking Ritalin (a stimulant) in the United Kingdom increased from 92,000 in 1997 to 786,000 in 2012.[51]

Beyond Anglo-American capitalism (sometimes referred to as the Anglocene or the Capitalocene to describe the origin of the Anthropocene), the benefits of efficacy and efficiency were not questioned by Stalinist states or by theocracies. The Anthropocene is therefore aptly named: it is the product of a deep human drive for performance, regardless of the political and economic system.

Optimisation also contains its own shield. It carries the idea of improvement without specifying what is being improved. Optimisation is thus neutral, and therefore beyond reproach. Optimisation may refer to the improvement of anti-spam filters to block out unwanted messages, or it might concern the improvement of the dissemination of computer viruses capable of bypassing the aforementioned filters.

Last, optimisation refers to a personal trajectory. It would correspond to a relative, perceptible, enviable, virtuous improvement. There is no shortage of famous quotations on this topic, a fact that is reflective of its widespread acceptance. For instance, André Gide said: "It is good to follow one's bent, so long as it leads upwards".[52] Similarly, in the Ethic, Spinoza does not praise the virtue of perfection, but rather that of perfecting: it is not a matter of achieving absolute perfection, but instead of experiencing the satisfaction of relative progress.[53] Let us cite also moralist Joseph Joubert: "The aim of an argument or discussion should not be victory, but progress". The inner satisfaction of progress, in whatever form, remains a driving force.

Our fascination for performance has in fact much more complex roots, and notably many cognitive biases. In particular, aspiration is a mobilising force in many spheres. Within the management sector, it is even considered key to stimulating collective commitment. However, this may result in a spiral of amplification. As such, upon reaching a certain stage, this mobilisation could also become self-justifying and may no longer maintain a link with the objective that was initially defined. There is no shortage of examples to back up this hypothesis. For instance, escalations in commitment are a familiar concept to psychologists: an individual or a group will not change course, even if the intermediate results are negative. The irreversibility of escalations in commitment by means of a ratchet effect is illustrated in various contexts, from the war in Vietnam to many failed construction projects, such as the costly, but never operational, Shoreham Nuclear Power Plant near New York.

If the race to optimisation is the result of a diffuse human impulse, it is difficult to measure its effects. By no means does a superficial analysis of human psychology provide sufficient grounds to associate optimisation and the Anthropocene. Let us now embark on a detour, and submit the link between optimisation and the Anthropocene to the test of History. The seminal work by Christophe Bonneuil and Jean-Baptiste Fressoz *The Anthropocene event* offers a number of telling examples.[35] It also highlights other mechanisms behind this race before us.

2.6 A MILITARY CATALYST

The over-exploitation of ecosystems within the Anthropocene does not occur in a linear fashion, but rather in successive stages. War is the main catalyst for optimisation. In fact, the word "engineer" appeared in the 16th century to describe builders of war machines. More recently, the efficiency of engines doubled and the automotive industry quadrupled its production output during the four years of the First World War. The link between industrial optimisation and the war effort is obvious: performance must be increased to counter the enemy. However, how can we explain the fact that the production of chemical fertilisers also increased during the war? All becomes clear when we note that synthetic nitrates are made from ammonia, which can also be used in the production of explosives. The first steps of the green revolution were by no means pastoral.

Incidentally, nitrate is one of the most iconic markers of the Anthropocene. This merits considerable reflection. Its abnormally high concentration in modern-day soils (alongside plutonium, another product of the military industry) is considered to be one of the clearer signatures of the global human footprint within the planetary stratigraphy. Some may raise the objection that nitrate, in its simple NO_3 formulation, is a commonly occurring natural element that is by no means anthropogenic. Here the resonance of the Anthropocene really becomes apparent; the production of nitrate brings together all of the ingredients of our time: ecological, historic, economic and geopolitical.

Indeed, the so-called Haber-Bosch process enables the conversion of nitrogen from the air into ammonia and then into nitrate. Because this enables the saturation of soils with nutritive elements, this process increases crop yield in the short term. The industrial production of synthetic fertilisers is incidentally one of the main causes of the drastic reduction in famine since the second half of the 20th century.[54] However, the synthesis of nitrate is energy-intensive: it takes 3 litres of oil to produce 1 kg of nitrogen fertiliser. In light of this fact, it is unsurprising to note that one of the world's leading producers of nitrogen fertiliser is Saudi Arabia.[55] It is once again evident that modern agriculture, using chemical fertilisers, is in fact still operating within a slash-and-burn culture: it is merely oil that is being burned rather than forests.

From a geo-biospherical perspective, the Haber-Bosch process represents a fundamental rupture in Earth cycles. We are enriching the ground with a product derived from the nitrogen in the air, regardless of the resulting consequences for organic material cycles. Together with ploughing, what results amounts to a form of desertification in the long term. We have created a new nitrogen cycle, on a planetary scale. We are operating in a similar way to some plants that, *through* bacterial symbioses, are capable of fixing nitrogen from the air. However, unlike these symbiotic plants, we consume a great deal of fossil energy, we pollute and we fuel desertification.

This marker of the Anthropocene is often overlooked by the media, with CO_2 stealing the limelight. However, it is of vital importance. Nitrate has clear demographic implications, with chemical fertilisers increasing crop yields.[56] The fact that there are about 8 billion humans in the world today is to a large extent down to the Haber-Bosch process. More precisely, it is estimated that of the total population of 7.4 billion humans in 2015, the equivalent of 3.5 billion individuals existed *as a result of* the indirect contribution made by chemical fertilisers to the production of human foods.[57, 58] In retrospect, it is perhaps surprising that Fritz Haber is not considered one of History's heroes, ranked alongside the likes of Alexander the Great or Joan of Arc, his contribution to humanity being significant, to say the least.

Alongside agriculture, the entire economy is optimised during times of war. While global trade has always existed, a qualitative leap forwards took place in the mid-20th century, with the proliferation of shipping containers, a physical marker of the globalisation of trade. So where is the link to war? Science historians Bonneuil and Fressoz recount the efforts made by McLean in convincing the Pentagon to allow him to take charge of the provision of food and supplies to soldiers in Vietnam.[35] This was realised, thanks to the first use of shipping containers. In order to optimise journeys from the United States to Vietnam, McLean established his own triangle trade by passing via Japan on the return journey, where he could load up on merchandise for export to the United States. This was the start of economic globalisation on a mass scale. The emergence of shipping containers would probably have occurred without the Vietnam War; however, this war was undoubtedly a catalyst for the advent of massive commercial trade, and its derivatives within the current digital economy. Where would Amazon or Alibaba be without shipping containers?

2.7 A RATCHET EFFECT

This historic analysis would be incomplete if we were to fail to mention the irreversible nature of increases in performance achieved during times of war. For example, France adopted Taylorism in its factories during the First World War. The war effort fully justified the optimisation of processes, since it did enable an increase in efficiency and an ability to respond to the urgency of the situation in the short term. However, with peace restored, factories retained the Taylorist model.[35] One of the key mechanisms of the Anthropocene is that human beings never backtrack on these developments. We fail to ask some pertinent questions: Is Taylorism justified in times of peace? And what of the proliferation of shipping containers? The mass production of nitrate? We proceed in successive increments, generally without a feasible way back, there being few arguments that speak out against optimisation. The Anthropocene is a one-way race to optimisation, in which we progress forward by way of a ratchet effect.

When all is said and done, why is this a problem? In light of the warnings outlined earlier, why shouldn't optimisation represent the very best of all possible solutions? We are going to have to move quickly, and we will certainly require the full scope of human intelligence in order to respond to the challenges presented by the Anthropocene. It could be argued that increases in productivity and competitiveness also constitute increases in agility. Thus, should we then increase our level of commitment yet further on the road to enhanced performance?

This rhetoric is very prevalent, and it is also in keeping with the enthusiasm generated by the possibility of building augmented humans or cyborgs.[54] This is also coherent with the plethora of superheroes churned out by Marvel and various other Hollywood studios. The next stage could in fact be much more global. Far from being consigned to science fiction, the possibility of bypassing all biospheric and geophysical intermediaries and feedback, gaining direct control of the climate by means of geo-engineering, is also being touted.[59]

In this regard, the excess CO_2 that we currently generate may be a so-called necessary evil if we are to make the leap to energy sobriety. We are in fact developing technologies with negative emissions, in particular *via* the sequestration of carbon in plants. However, in theory, this would require that an area two to three times the surface area of India be dedicated exclusively to this task in order to achieve carbon neutrality today.[60, 61] Moreover, this would only be a temporary measure, since the exploitation of these trees for human purposes or their final decomposition would release the sequestered CO_2 at the end of the cycle: a reduction in the quantity of CO_2 only takes place during the growth phase of the trees. Incidentally, old forests are now in a state of balance, and for this reason they no longer sequester CO_2. In order to reduce the quantity of atmospheric CO_2 using plants, it would therefore be necessary to plant trees with a very long lifespan, and this would only provide us with a reprieve lasting for a few decades.

Other geo-engineering approaches are being proposed as a means of reducing atmospheric CO_2 concentration. Discharging iron sulphate into the ocean, as part of a process known as *fertilisation of the ocean*, is said to stimulate the proliferation of vegetal plankton. The hope here is that the associated photosynthetic activity would reduce atmospheric CO_2 in mineral form, for example, in the skeletons of diatoms that then settle on the ocean bed.[62] However, the assessment of the amount of CO_2 actually sequestered by this process remains highly debated.[63] Moreover, the oceanic biodiversity would inevitably be strongly affected. Other proposals have also been put forward. For example, the dispersion of sulphate aerosols into the atmosphere, as occurs during the eruption of a volcano, could screen out the sun's rays and cool the planet.[64]

But where does this planetary control stop? Could we ever turn back? Have we correctly assessed the risks of such a venture? It should be noted that geo-engineering technologies are subject to international moratoria: they involve all of the Earth's biospheric and atmospheric cycles and thus have a true extinction potential. On a more fundamental level, are we not simply burying the thermometer rather than facing up to the initial cause of the climate crisis? Very much in keeping with this insinuation is the fact that the oil industry lobbies are often keen on geo-engineering. They even block attempts to pass international regulations promoting a precautionary approach in this area, in order to justify the existence and development of their industry.[65] Given our past trajectory and the challenges ahead, we may well take the plunge into geo-engineering, despite these warnings. This final ratchet could take us into the world of total control and optimisation. A high-risk project!

Besides war and its performance drivers, finance may just be the beating heart of the ratchet effect optimisation prevalent in contemporary societies. Indeed, the deregulations of neoliberalism, conceived in the 1950s by Milton Friedman and Friedrich Hayek (Mont Pelerin Society) and implemented on a large scale in the 1980s by Ronald Reagan and Margaret Thatcher, altered the means by which companies and states are funded.[66] In particular, the rate of borrowing by states is currently dependent on indicators issued by rating agencies, which become increasingly more famous with each successive crisis. The posting of such indicators itself promotes the objective of enhanced performance, by means of successive increments and within a loop of self-amplification, without turning back. Within such a world, the objectives to be achieved are no longer real. They are crystallised within a rating, the maintenance of which implies a curve of continual growth. This rating is self-justifying and replaces other, more tangible objectives, such as the well-being of citizens or the preservation of ecosystems. Ultimately, this brings to mind Zen teaching: if you have achieved your goals, you have also missed everything else.

Thus, our trajectory is that of growth and optimisation, in successive increments, without any foreseeable return route. However, we could reach some limits today.

As the Frankfurt School philosophers Max Horkheimer and Theodor Adorno wrote, "the power of industrial society is imprinted on people once and for all".[67] Today, the rationality of progress catalysed by each successive war has become an irrationality of perpetual violence. This is best evidenced by the destruction of our earthly habitat and the current epidemic of burn-outs.

We have a schizophrenic relationship with performance. For a few top billings, how much frustration does it cause at work, in sporting competitions or at school? Happiness is a form of mental conditioning, with the promise of a world that is richer in sensation, knowledge and adventure. However, frustration and deception are also compounded. For example, between 1985 and 2017, the number of suicides quadrupled in South Korea, even though the country emerged from a dictatorship and was booming economically.[54] Burn-out and systemic unemployment appear to be indicators of a society that is overheated, and therefore not socially sustainable. Our ideological belief in performance clashes with what it brings forth in the real world.

How to reclaim progress? What rationality of peace could we imagine? What alternative to continuous optimisation could we propose? This is the main purpose of this book. Before exploring the contours of this third way, let us dig into the complexity of the world. An appraisal of past civilisations can be enlightening. They grew in successive stages but they also disappeared abruptly in part due to the forcing of their resources. From the Mayans to the Khmers, when a population grows, so too do its needs, and the exhaustion of resources follows suit.[68, 69] The Anthropocene invites us to better understand this feedback.[70]

2.8 COMPLEX FEEDBACK

Up until now, the race to optimisation has somewhat widened the gap between what is *best* for humans and what is *best* for Earth. We find ourselves in the same situation as Raphaël de Valentin, who was granted successive wishes at the cost of shrinking his "wild ass's skin".

In contrast to Balzac's novel, such feedback rarely exhibits linear behaviour: some unexpected accelerations may occur. For instance, the Atlantic cod has been an abundant resource during the five centuries of its fishing. Yet today it is an endangered species. A simple explanation lies in the period of overfishing during the 1960s and 1970s. To address this decline, a moratorium was finally enforced in 1992. What happened next? The adult cod is a predator of smaller fish, such as herring. With the reduced number of cod, small fishes such as these could proliferate and become omnipresent predators for juvenile cods. In the end, the collapse of cods accelerated, in spite of the moratorium. Thankfully, it seems that the first signs of a restored balance are in sight, but it was a close call.[71] This example illustrates how the decline of one species can disturb an entire ecosystem, even converting prey into predator. More generally, human impacts on any ecosystem are and will continue to be the cause of multiple instances of feedback. We are entering into an epoch in which ecosystems will shift away from the norm, and in unpredictable ways.

The human brain alone is too limited to grasp such complexity and explore the evolutions of these different scenarios. This is one of the Achilles's heels of optimisation: it is easy to know what to optimise and how to do it when the problem is clearly defined. Optimisation is adapted to closed systems. Political and economic decision-makers deal with questions that have multiple dimensions, and for which the optimisation of one element does not necessarily have a beneficial effect for the system as a whole. If one adds the strong uncertainties to come, our system becomes very open, and optimisation becomes less and less relevant. The example of the biofuel sector offers a prime example of this poorly anticipated feedback.[72] Here are the salient points.

It was, to some extent, the impact of Hurricane Katrina on the oil refineries of New Orleans that drove the US administration to vote in the Energy Independence and Security Act (EISA) in 2007. This act advocates a better control of the supply of oil, notably through the development of a biofuel industry. George W. Bush had withdrawn from the Kyoto protocol at the start of his term in office, yet he used the climate crisis argument to justify this new direction. The United States thus began the mass production of bioethanol, followed closely by the European Union, which voted in a law equivalent to the EISA in 2009. These laws served as green lights to palm oil–producing countries. In fact, each fully grown oil palm produces 25 kg of oil-filled fruit every fortnight, the highest output among oil-producing crops. This crop ticks the boxes of both efficacy (achieving the objective of oil supply control) and efficiency (an excellent ratio between the number of plants per hectare and the quantity of oil produced). Indonesia consequently decided to convert 52,000 km^2 of forestland for the industrial production of palm oil. This is equivalent to the surface area of Costa Rica.

In 2008, that is, one year after the American vote and one year prior to the European vote, an article in the very prominent *Science* journal[73] revealed that such a conversion of land for biofuel production could double the production of greenhouse gases. How is this so? If we were to only take into account the type of fuel, burning bioethanol would certainly reduce the quantity of greenhouse gas emitted when compared to fossil fuels. However, if we take into account the entire associated agricultural cycle, and in particular, the lands to be converted and the associated deforestation, then the conversion of agricultural land for biofuel use becomes a heresy. This warning went unheeded and today, backed up by corresponding figures, it must be recognised that the biofuels industry has indeed led to an increase in greenhouse gas emissions.[72]

Other collateral effects of biofuels have been even more poorly anticipated, or wilfully ignored. Massive deforestation in Malaysia and Indonesia is substantially altering habitats. Biodiversity is collapsing in these countries, the iconic example being the endangered Bornean orangutan. Some of the cultures and traditions of Indigenous peoples are in jeopardy. It is worth mentioning that Indigenous peoples still represent a great source of cultural wealth: they number 370 million to 500 million individuals, speaking 4,000 languages, and distributed throughout more than 70 countries.[8] Deforestation is also generating widespread air pollution in the region: in Indonesia, up to 120,000 huge forest fires were visible using satellite imagery at a given moment in 2015! Although Beijing and Delhi often steal the limelight, Jakarta is often ranked among the most polluted capital cities in the world, mainly on account of the indirect effect of deforestation on the quality of the air.

The United States consolidated its own corn-based biofuels sector, and the production of shale gas soared in the 2010s, in perfect coherence with the initial economic demand of the EISA. As a consequence, the demand for palm oil for biofuels dried up, but other needs emerged. The majority (two-thirds) of palm oil produced today is in fact used for food, mainly for highly processed food products such as potato crisps or spreads.[74] Not only are these products of practically no nutritional value, but the saturated fatty acids that they contain can even be harmful to one's health. Moreover, it is worth noting that the inclusion of palm oil in food products is often concealed behind multiple different names: there exist 25 variants.[74] The palm oil industry is therefore a cultural, nutritional, medical and ecological impasse.

Thus, what direction should we take moving forward? While the problem has been addressed with better effect in some countries, the damage has been done: there is no sufficiently quick solution when it comes to putting a stop to such an industry and its investments. The United States, followed closely by the European Union, actually decided on behalf of the entire world that the biofuels' industry was the way forward. The countries concerned invested structural funds in this economy and have reached a position from which they cannot easily turn back. By way of an illustration of these delay effects, and of the associated denial, Indonesia is thus set to increase its use of palm oil for its own biofuels. Even worse, new players are entering the market, despite the above observations. Ambitious programmes for the development of palm oil are appearing in Sierra Leone, at a crossroads between neocolonialism, the expropriation of land and violated human rights.[75] Europe's backtracking regarding palm oil would seem, at best, to be a late realisation of past errors.

In addition to the cultivation of oil palm, there are many other examples of political decisions that ignore the multiple negative feedback effects. For instance, consider the support for the import of Brazilian soya for European cattle farming, which promotes land conversion for agriculture and is thus indirectly responsible for the ignition of Amazonian forest fires. With France being the third-largest importer of soya bean meal from Brazil, President Emmanuel Macron has had to acknowledge "a degree of complicity" in the Amazonian forest fires on the part of the French State.[76]

So, is our choice to ignore the feedback within the economic sphere simply the result of our being blinded by the potential short-term gains? This hypothesis is unsatisfactory as history shows that we ignore feedback even when the economic challenges are less direct. Let us consider the case of humanitarian aid.

Dichlorodiphenyltrichloroethane (DDT) is an insecticide that was used during and after the Second World War to combat malaria contracted by soldiers. DDT is therefore first and foremost a molecule that is supposed to indirectly support the performance of the army, thus joining the vast collection of markers of war in the Anthropocene. This molecule was at the heart of several controversies. Most notably, in the famous book by Rachel Carson, *Silent Spring*,[77] it was held responsible for the reduction in the number of birds in the United States. This resulted in it being banned in several western countries. The other feedback effects on the trophic chains of ecosystems have been very poorly anticipated.

For instance, a malaria epidemic broke out in Borneo in the early 1950s. The World Health Organization (WHO) decided to spray DDT in order to kill mosquitoes, drivers of the epidemic. A cascade of secondary effects was then triggered. For example, cats became poisoned and died as a result of rubbing up against the walls of the houses and then licking their bodies. With a reduction in the number of predators, rat numbers soared. While rats do not transmit malaria, they are carriers of diseases that are just as serious, such as typhus or the sylvatic plague. In response to this risk, dozens of cats were even parachuted into one village![78]

Might the bureaucracy of public services be a victim of its own complexity, thus preventing it from predicting feedback effects? Let us consider the private sector and its interpretation of humanitarian aid. With the objective of eradicating specific diseases such as AIDS or malaria in Africa, the Bill & Melinda Gates Foundation has provided huge financial support for certain initiatives in the countries affected by these diseases. While this objective is commendable in principle, it has also given rise to significant bias within the health systems in these countries, in particular by weakening their capacity to combat other illnesses. It thus becomes difficult to recruit a generalist nurse if a foundation decides to triple the salary of nurses who specialise in the fight against AIDS. In hindsight, such a targeted and simplistic strategy could only create a bias for local healthcare professionals. Incidentally, public/private partnerships in healthcare are often under fire nowadays due to the counterproductive distorting effects that they generate.[79, 80]

There exist a number of examples of poor decisions that are very simple to understand, but whereby the feedback has not been properly taken into account, to the point that it weakens the economy or the health of the populations. In summary, because feedback is counterintuitive and poorly predictable, optimising in an open system only presents risks.

2.9 MODELS AND THEIR LIMITATIONS

In the vast majority of cases, intuition, ideology and emotion serve as a guide for decision-makers in choosing an optimal solution to a complex problem. In addition to these non-rational approaches, we have seen that computer modelling has occasionally been welcomed into the decision-makers' circle to help formalise predictions, and thus guide future choices in a more rational fashion.[36, 39, 47, 44] Might modelling be a tool that should be added on a wider scale in order to overcome the challenges of the Anthropocene? Before going any further, we must accurately define what modelling entails.

The theory of complex systems offers a mathematical definition of complexity. A complex system is a group of elements that interact with one another according to simple laws. For example, a pile of sand is a complex system: the grains of sand interact with each other according to the laws of gravity and friction. In the same way, a cloud is a complex system in which suspended water particles interact with one another. All living organisms are complex systems, as they are the fruit of interactions between cells, resulting in the formation of tissues and organs. These are actually multi-scale complex systems, since, akin to Russian dolls, living organisms are the result of interactions between molecules, between cells, between tissues, between organs, and finally, between organisms within the population.

However, a complex system is not a complicated system. Watches and mobile phones are complicated systems. What is the difference? Within a complex system, local interactions are sufficient to explain its form or global behaviour. There is no need for a leader or architect because the system is essentially self-organised. In this regard, an anthill is a complex system: the queen ant is not the architect of the galleries of the anthill; interactions between ants account for the ultimate diameter and organisation of the galleries. Within a watch or a mobile telephone, there are many interactions, but they are not in evolution. An architect has decided to position them in accordance with a defined plan. A complex system operates on a bottom-up and a top-down basis, thus including feedback, whereas a complicated system only operates on a top-down basis.

It is here that we see the merit of the complex system approach in integrating and better understanding the feedback within a system such as planet Earth. For instance, the daily weather forecast is based on computer models that draw on the theory of complex systems. The World3 model, cited earlier, is also a typical example of a computational tool built using the complex systems approach.[36] Beyond predictions, this type of model has an essential virtue: documenting the feedback loops provides a means for understanding the structure and dynamics of the system, an essential step for its eventual correction. In other words, before "changing the system", we must understand it. As the authors of the Meadows report write: "When we, system dynamicists, see a pattern persist in many parts of a system over long periods, we assume that it has causes embedded in the feedback loop structure of the system. Running the same system harder or faster will not change the pattern as long as the structure is not revised".

While, in theory, computer modelling may serve as a guide for political and economic decision-makers, it must remain humble: a model could be considered genuinely predictive if all interactions were described and included, and this would in turn require that the proposed solutions were acceptable. We are still a long way off when it comes to the Earth, ecosystems or humans.[81]

How can the even partial consideration of the complexity of systems, and the associated feedback, help us to understand the tropism for the optimisation of human beings? The predictions produced by models are often counterintuitive, or even indefensible. For instance, subsidies for livestock farming in Crete, intended to support local agriculture, instead guarantee the permanent desertification of the island in the medium term. In order for Crete to be inhabitable in 50 years' time, it may be necessary to tax farming today so as to afford plant cover the chance to reform. Here we can sense the contradiction between the desire for what is *better* for humans, and the predictions of models that may stand in opposition to this desire. Furthermore, even if a model is mathematically correct, it is built of human hands and as such it carries within it this human imprint. In particular, a model can be easily biased in order to justify ludicrous decisions made on scientific grounds. The example of overfishing provides the perfect illustration of such ambiguity.

Increasingly aggressive fishing techniques enable an increase in yield and, in a sense, the optimisation of fishing voyages. Nevertheless, for a long time these tools have also raised concerns of a threat to the fish population's capacity to regenerate. According to FAO, 58% of fish stocks is being exploited to its maximum potential, 31% is being over-exploited and just 10% is under-exploited.[82] Fishing quotas have been discussed, which have the effect of curbing the levels of performance desired by fishermen and the respective states. However, a pseudo-systemic approach has been adopted in the development of these quotas. It was indeed proposed that when too many adult fish are left in the sea, the young fish do not have sufficient resources to develop. It was therefore not a matter of defining an absolute quota, but rather a relative quota that can be integrated into the fishing catch/effort ratio. This serves as an indicator of the maximum quota that cannot be exceeded if we are to preserve this resource (*maximum sustainable yield*), but one that enables us to move beyond arbitrary quotas. This means of formulating quotas could make sense if the population of a species of fish were disconnected from other species within its environment. Since this is obviously not the case, fish reserves have declined throughout the world.[83]

This example illustrates how the complete description of a complex system is required in order to understand its dynamics and function. It also illustrates the ambivalence of the systemic approach that can be used to justify absurd political decisions. Without being too cynical, it is difficult to imagine that those responsible for conceiving the concept of *maximum sustainable yield* were not conscious of the irreversible and non-linear effects on fish populations within an open system such as the ocean. We can clearly see how the objective of optimising profits in the short term, consistent with the principle of the freedom of the seas, which is essential for military fleets,[35] took precedence over a genuinely systemic vision for fishing.

Our desire for efficacy and efficiency, complete with its internal driving forces, its military catalysts and its systemic contradictions, appears to be unyielding. The Anthropocene is indeed the age of performance. In turn, we are beginning to see the effects of a degraded environment on our lives. Instead of taking this into consideration, our reactions feed amplification loops. The age of performance is also the age of feedback. Faced with the prospect of a systemic tipping point during the 21st century, a question becomes more and more pressing: What should we do? Confronted with such a total, plural and synchronous revolution, there exists a wide range of attitudes, ranging from indifference to denial, from sideration to the urgency of activism. We have seen that the tropism of optimisation has served as a catalyst for the Anthropocene, and that solutions of appropriate complexity require a new approach. What kind of world could we be heading for?

3 Which Third Way?

If the lessons of complexity push us to more humility, let us first be certain to consider all courses of action, from the feeblest to the most ambitious. Let us now explore the different responses to the challenges of the Anthropocene. While we are agonising about the future, Elisabeth Kübler–Ross's five emotional stages[84] (denial, anger, bargaining, depression, acceptance) could help us to put them in order. In this chapter, we will go a little deeper into the challenges ahead. As a warning, the following chapters might seem "eco-anxious". My aim is rather to scrutinise our world of performance, without making too many concessions. This is a prerequisite before we can discuss an inspiring new path in the following chapters.

3.1 BUSINESS AS USUAL

Why not just adopt the *business as usual* approach, since the system appears to be self-regulating? Within that scenario, fears would be exaggerated. "If there is no solution, it's that there is no problem" (Les Shadoks). There actually exist intangible markers that can ease, and even seemingly justify this denial. The rate of infant mortality has never been so low. Famines have never been so rare. When it comes to the environmental crisis, it would appear that we are reaching a turning point, without making a great deal of effort. The development of renewable energies and the blossoming of an environmental consciousness should even provide new economic opportunities for the future. Therefore, is the status quo a viable solution?

As indicated earlier, it is all a matter of timing. If the *Titanic* had turned earlier upon approaching the iceberg, it would not be recorded in the history books. Are we moving fast enough to provoke change? Given the time required for a forest to regenerate, for soil to be replenished, for the climate to return to the norm of the Holocene or for new species to emerge, it would take extraordinary measures to reverse the current trend. For instance, CO_2 remains in the atmosphere for a hundred years once emitted.[8] It has also been highlighted earlier that the oil industries have not announced any intention to ramp down their production. On the contrary, in 2019, British Petroleum (BP), Shell, ExxonMobil, Chevron and Total Energies invested a total of 201 million dollars in lobbying, with a view to delaying or blocking policies that promote the battle against the climate crisis.[85] On a more fundamental level, we have shifted the burden of human advancement onto ecosystems. Today, as is evident in the collapse of biodiversity, ecosystems simply cannot endure it any longer.[47]

Ironically, the strongest advocates of business as usual often demonstrate selective ignorance, or perhaps even techno-schizophrenia: either the science is wrong and there is nothing to fear, or we face a genuine challenge; however, science and technology will respond to the current challenges in the same way as they have done in the past. Supporters of the status quo often see environmental activists as dangerous extremists. However, business as usual is by far the most radical scenario, as the associated

DOI: 10.1201/9781003510918-3

project is literally suicidal. Unfortunately, radicalism, entirely occamian on account of its parsimony, is often more engaging than complexity, being the only solution to offer a perspective that is more easily comprehensible to the human brain. Doing nothing because science is wrong or because science will solve everything provides a means by which to avoid asking ourselves too many questions. Yet if you arrived on Earth from another planet and were told that a single species was solely capable of bringing about a mass biological extinction, as never before seen since the appearance of life on Earth four billion years ago, would you not be concerned? How can we explain away the fact that the warning associated with the environmental crisis has so little effect?

3.2 WHY ARE WE PASSIVE?

The formalisation of the Anthropocene reveals a temporal dissonance: the synchrony of many crises (climate, biodiversity, pollutions) versus the diachrony of our perception and our responses. Long before Hannah Arendt[86] or Stanley Milgram,[87] history has shown that humans are capable of ignoring warnings, of remaining passive in the face of massacre or violence, then justifying their passivity *a posteriori*. Here it must be acknowledged that we do have an excuse: our internal weaknesses.

Our psychological biases are many, and they provide part of the answer to the question as to our inaction when faced with ongoing and pending ecological disaster.[88] Consequently, Jared Diamond, author of the famous essay "Collapse", cites a survey on perceived fear among populations living downstream from a hydroelectric dam. Intuition would suggest that people located further from the dam would have less fear than those living at the foot of the dam. The results of the survey show the exact opposite. This can be easily explained by psychological denial: in order to be able to live peacefully, our brain forgets the dam and the possibility of it bursting.[69]

Likewise, the anchoring effect – the common human tendency to rely too heavily on the first piece of information given – urges us to avoid changing opinion so easily. The aforementioned case of escalations in commitment shows how our natural bias will tend towards conservatism. "In for a penny, in for a pound", even if the facts show our fixation to be erroneous.

American social psychologist Leon Festinger studied the surprising responses of humans when their deepest convictions are challenged, even going so far as to speak of *cognitive dissonance*.[89] For example, a group of Americans was persuaded that the end of the world would occur on 21 December 1954 and that extraterrestrials would save them by abducting them at midnight on that date. However, at midnight, nothing happened. At 00:05, the group found a clock showing the time as 23:55 and thus decided that it was not yet midnight. At 2:00 am, a young boy remembered that he had to return home (otherwise, his parents would call the police). The group did not join him in returning home, and told him that his sacrifice would enable the group to be saved. Approaching 5:00 am, the group convinced itself that it had succeeded in preventing the end of the world.[66, 90] This stubbornness extends beyond abusive cult practices: when reality does not correspond to their convictions, humans prefer to alter reality rather than change their opinion.

This is not an intellectual limitation, but rather a cognitive bias. The most educated people are also the most reluctant to change their opinions. Regularly engaging in debate, they are not short of arguments by which to achieve the desired response, in keeping with their ideology, even if this response is incorrect.[91, 92] Moreover, this bias can be verified by means of experiment. By comparing the response of a group of people to the results of a virtual study on the effectiveness of a cream or on the effectiveness of gun control, using exactly the same figures, Dan Kahan and his co-authors demonstrate this ideological bias. In the first instance, more educated persons avoid the pitfalls of intuition and provide the correct response. For example, the cream reduces irritations in a larger number of patients, but as a percentage of the total group, this cream actually increases the proportion of patients presenting with the irritations. However, these same, more educated people provide an incorrect response when it comes to gun control, as a result of ideological conviction, even though the mathematics are identical.[93]

We could look further back in time. The theory of heliocentrism (that of the Earth orbiting around the sun), as opposed to geocentrism, was opposed by the Catholic Church up until 1741, some 54 years after the formulation of the theory of gravity by Isaac Newton (1687), 130 years after the observations by Galileo Galilei (1611), 132 years after the definition of the elliptical orbits of the planets around the sun by Johannes Kepler (1609) and 228 years after the heliocentric proposal by Nicolaus Copernicus (1513). What is true in the educated religious world is also true in the educated scientific world. As Max Planck put it: "A new scientific truth does not triumph by convincing its opponents and making them see the light, but rather because its opponents eventually die, and a new generation grows up that is familiar with it".[94] Thus, it would appear that we are, by default, conservative.

These examples of psychological bias suggest that all of the rationality and teaching have only a minor impact on our actions: we know – occasionally we know a great deal – but we very rarely react accordingly. From that standpoint, we are therefore all responsible: our actions are not strictly consistent with our understanding of the global problem, which can indeed be overwhelming. This is what Jeremy Rifkin calls *cold evil*.[95]

Are we failing to respond to the warning because it is recent? This is unlikely. It has already been issued several times in the past. The alert was sounded regarding the exhaustion of carbon resources in the United Kingdom in the mid-19th century,[96] the end of oil resources began being publicised in the United States from 1921 onwards[97] and the end of mineral resources from 1945.[98] We could even go back to Plato in Critias, who already observed the harmful effects of deforestation (albeit without attributing it to human activity). We know that we are approaching the exhaustion of resources and that we are altering the balance of the Earth. Yet this awareness is not accompanied by relevant action. Ironically, we have never released as much CO_2 since the beginning of the industrial era as we have since the creation of the IPCC in 1988, intended to curb the rise in the atmospheric CO_2 level!

Bonneuil and Fressoz speak of the *Agnotocene*[35] as a means of qualifying this era in which our perception of reality is orthogonal to the physical reality that surrounds us. It is not that there exists a lack of information, never have we had so much! From this perspective, the Agnotocene marks the failure of the predictions of modellers, whose objective is not solely to be realistic, but to also have an impact at the moment in time at which the models are formulated. Although we are aware of our reality, we make the deliberate choice to avert our eyes, even with the end of resources becoming palpable, the widespread pollution of our land a tangible prospect and the collapse of biodiversity an established fact.

The coordinated multilateral response to the "hole" in the ozone layer based on scientific evidence[37] or the fact that more than 195 states adopted the Paris Agreement on climate change in 2016 suggest that the *business as usual* option has been abandoned, at least officially, by most countries. The scenarios of the IPCC provide a very clear indication as to the number of years it would take for us to reverse the trend when it comes to climate. For example, by following the very optimistic RCP2.6 scenario, we could achieve carbon neutrality by 2050, thanks to a global transformation of energy production, and could thus limit global warming to below 1.7°C.[99] Furthermore, the amplification of local forms of resistance shows change is definitely afoot. The emergence and rapid popularisation of the term *Anthropocene* are likewise signs of a turnaround. That said, we are going to have to find ways to address our culture of indifference, and therefore perhaps also our veneration of optimisation.

3.3 ACTIVISM

With business as usual, we were in the comfort of denial. This is not a viable option, so we naturally enter into anger. The field of political ecology emerged in the 1960s[77] and has not ceased to evolve since, from Greenpeace to Extinction Rebellion, via the numerous climate demonstrations organised by younger generations. Far from being anecdotal, this stage is essential to change. In light of this, it is important to highlight one figure: 3.5%. This is the percentage of the total population that is believed to be the threshold beyond which a non-violent movement brings about a radical societal change, according to an exhaustive study of the non-violent demonstrations of the 20th century.[100] In this context, angry protests in the street are actually symptoms of a social fabric that is still alive, breathing out hope of a transformation. It is, in fact, the apathy of some democratic countries that should be the real cause of concern.

Because, in addressing anger, we are entering into sensitive and emotional territory, I would like to highlight how art has taken possession of this issue. Contemporary art in particular has become a form of complete philosophy, which questions humanity in its multiple dimensions, and as such finds in the Anthropocene a prolific field of play.[101, 102] Below is a subjective selection of examples through my own encounters with artists, events and exhibitions.

Performing arts, situated at the boundary between the event and the theatre piece, has been and continues to be a medium through which to question our place on Earth. For instance, at the University of California, Berkeley, the *Ant Farm* collective (1968–1979) constructed *Clean AirPods* and stirred up the crowds while wearing gas masks, imploring them to take refuge in an inflatable structure, warning that the air would soon become unbreathable. On a similar vein, with its famous *Cadillac Ranch*, a line-up of Cadillacs "planted" in the countryside, the collective engaged in criticism of the automobile society and its underlying absurdity.[103] Here we find links with critique of consumer society expressed by Andy Warhol and other protagonists of the pop art movement.[104] Famous films such as Richard Fleischer's *Soylent Green* also expressed similar outrage.

Another, perhaps more counterintuitive, form of activism is manifested in stupor and terror, the two ingredients of the sublime according to Edmund Burke.[105] Lucretius, before him, established links between stupefaction and the sublime: "all that stuns the soul, all that imprints a feeling of terror, leads to the sublime".[106] In fact, as French geographer Nathalie Blanc expressed it in her lecture on Aesthetics and the Anthropocene, "the sublime rests on the sentiment of our own insignificance when faced with a vast, distant nature, suddenly demonstrating its omnipotence".[107] Here we find ourselves at the very heart of the Anthropocene. The romantic beauty of the collapse has always been a source of inspiration for artists. The stupefaction associated with the Anthropocene provides a rich and multifaceted source of inspiration. Driven to the extreme, the Anthropocene can generate a form of negative pleasure, a certain brand of enjoyment found in pain. For example, the years of recession in Detroit have attracted photographers from all around the world and this city has even been designated as the capital of *ruin porn*. The documentary *Our Daily Bread* (Nikolaus Geyrhalter), elegantly articulates the sublime and activism though a clinical description of intensive agriculture.

Artists also express their anger through humour, a medium that engages the public via an alternative route. Alongside from pop art, contemporary art exhibitions abound with works highlighting our delay in the face of the immensity of the coming challenges. For example, the Boston Institute of Contemporary Art presented a distinctive work in 2010 (*Tar Bears* by the artist Charles LeDray): a plush bear mired in an oily ooze, surrounded by pieces of sugar, as if to sweeten the scene.

Somewhat more Zen, land art, in transforming a natural space into an artistic space through the alignment of stones or branches (e.g. in the works of Nils Udo), demonstrates the human being's control over the *in situ* environment, nature no longer being an object that is opposed to nature. The Anthropocene is indeed probably the final nail in the coffin of the dualism between nature and culture.[15] At this point, reference could be made to the paintings of Yang Yong Liang (*Industrial Pollution*), presenting the traditional Chinese painting styles of the Song and Ming Dynasties, with the notable difference that upon closer inspection, one can find that mountains are entirely made of tall buildings, hosting high voltage power lines and cranes gnawing away at what is now a cultural landscape. Conversely, one might say, photographer Peter Steinhauer likens high-rise buildings wrapped in coloured synthetic textiles during their construction to gigantic cocoons. It is worth noting that this holistic vision does not mask the genuine diversity of natures and cultures: instead, fusions of culture and nature give rise to new local complexities.[108]

With the question of the Anthropocene, art and sciences are more and more intimately associated. For example, Pinar Yoldas, with her *e-plasticeptors* (*an ecosystem of excess*), presented viscera in flasks of liquid, in the same way as biological tissues are preserved in formaldehyde. In a prospective scenario, these specimens might reveal how ocean-dwelling animals have found means of surviving massive plastic pollution thanks to symbiotic bacteria amenable to digesting this synthetic material. This interdisciplinarity also feeds geopolitical questions. For instance, at the Museum of Contemporary Art in Tallinn (Estonia), an installation presented photos of plant seed banks adapted to suit the arid climates of war-ravaged Aleppo, and compared them with the trauma of seeing a cadaver (Denès Farkas, *How-to-calm-yourself-after-seeing-a-dead-body Techniques* 2017).

Artists also often highlight the desire for the preservation and restoration of Earth. For example, in 2003 artist Maarten Vanden Eynde photographed himself plastering over the cracks in some dried-out earth, as if to repair it. This desire is also expressed in community projects with local residents, in particular within the *Seeds of change* project, a kind of green guerrilla war aiming to demonstrate the legitimate right of endogenous plants to live within their territory, however urban this territory might be. However, if anger is mobilising, then have we not already taken action and commenced the bargaining phase?

3.4 SUSTAINABLE DEVELOPMENT

The framework for action might actually be quite simple: we must reconcile what is *best* for human beings and what is *best* for the Earth. One response to this proposition is sustainable development, the stated purpose of which is to "respond to the needs of the present without compromising the capacity of future generations to respond to their own needs"[109]: recycling waste, making the transition to 100% renewable energy, establishing a pesticide-free system of agriculture, etc. Here we have a unifying project, and one that lists a consideration of the environmental problem among its merits.

However, it still remains ambiguous. The word *sustainable*, which expresses the need to take environmental constraints into consideration in order to live, contents the world of academia and NGOs. The word *development* satisfies governments and the private sector, by perpetuating the myth of eternal growth.[110] When George Bush Senior said, "*the American lifestyle is non-negotiable*", he was not opposed to sustainable development, in fact it was quite the contrary. Likewise, a report by the international Energy Transitions Commission (ETC) plans to reduce the consumption of CO_2 by half by 2040 thanks to huge investments in renewable energies, while still reconciling economic development.[111] For some, the climate crisis also generates business opportunities. It is not surprising to discover among the members of the ETC, industrial companies such as Shell and banks such as HSBC, perfectly illustrating the multiple contradictions of sustainable development. Sustainable development is at best a vague and consensual notion.[112] At worst, it is a dangerous oxymoron.[113] Sustainable development would imply the maintenance and regeneration of resources. However, our economy is far from being circular. Sustainable development is therefore only a fantasy today.

The management of plastics illustrates how sustainable development has even become a privilege of rich countries, subsidised by the ecological debts of poor countries. Between 1950 and 2015, we produced 8.3 billion tonnes of plastics. Of this phenomenal quantity, in billions of tonnes, 2.5 have remained intact and are still used today, 4.9 have been discarded, 0.8 have been incinerated and just 0.1 have been recycled and are still in use today.[114] An investigation conducted by *The Guardian* revealed how so-called recycled plastics actually end up at all four corners of the Earth. For instance, in 2018 the United States exported 68,000 containers (1 million tonnes) of plastics to poor countries such as Bangladesh, Laos, Ethiopia and Senegal, all of which have lax environmental regulations and are already struggling to deal with their own plastic waste.[115] In other words, the recycling of plastic can also power an economy that is far from green or virtuous. When it comes to sustainable development, we have not always managed to make it out of the denial phase after all.

The scenario of the sustainable development of our civilisation could, in theory, be credible if we had even more abundant resources, a great deal of time and a mentality capable of quickly abandoning consumerism. As mentioned earlier, only a minority of the humans on the planet are aware of this change in paradigm, and an even smaller minority are acting accordingly. The new cities of Asia illustrate this perfectly: the trio of motorway – skyscraper – mall is more than ever the objective sign of development today.

Considering the notion of sustainability is more engaging. For example, 400 European towns subscribed to the Zero Waste Network.[116] The roll-out of positive energy buildings, that is, buildings that produce more energy than they consume, is no longer a fad.[117] The development and progressive integration of these approaches to form a coherent whole, an authentic model for future society, represents a genuine glimmer of hope.[118] This prospect certainly explains the public success of documentaries such as *Demain (Tomorrow)*.[119] It is even already embodied in some pioneering companies, such as the ecologically minded envelope manufacturer Pocheco (near Lille, in France).

The balance between economic development and sustainability, however, calls for great vigilance. For example, to move beyond the age of fossil fuels, the all-electrical and all-electronic technologies require various minerals (e.g. for the design of batteries or electronic chips). However, their extraction is extremely polluting and may not be acceptable or viable in the future. It also requires a significant amount of energy (7% to 8% of global energy[120]). Cleaner sources of energy exist. These are often small, decentralised and repairable biomimetic units, with few rare minerals. Examples include silicon photovoltaic panels, and their possible future organic version based on electroluminescent micro-algae,[121] which mimic leaves, small windmills made of wood or recycled materials, whereby the aerodynamics bring to mind maple fruits, or even snake-like articulated tubes that recover energy from the waves with a view to converting it into electricity.[122] Our energy salvation could instead come from simpler and more local *low-tech* solutions that consume fewer rare elements, are repairable and thus are more sustainable.[123] The catalogue of options is thus heterogeneous, and we have reason to be optimistic that these could work as part of an integrated system in the medium term.[118]

One final red flag concerns the social aspect of sustainable development. At whom is this new economic model addressed? Is it a form of stealth gentrification? What is the average income of inhabitants of green towns? What types of job do the residents have? Are these initiatives without negative impact on other territories? If Jennifer Aniston eats an organic quinoa salad produced on the hills of San Francisco, she will indirectly increase the price of quinoa in Bolivia, depriving masses of people of a complete, healthy and ancient food, located thousands of kilometres from Coit tower.

Here we see that sustainable development has a complexity of its very own. The example of quinoa illustrates the need to adjoin the practices of sustainable development to binding international regulations. In particular, this would involve acknowledging the special role of farmers in assuring the survival of humanity. Following the model of cultural exception, this could take the form of an international agricultural exception: citizens and farmers at the heart of a system where staple foods would be taken out of the market to ensure fair remuneration for producers and priority given to the ecological and nutritional quality of products.[124]

This is far from being utopian: the ambitious Havana Charter, signed on 24 March 1948 by 53 states but never ratified, laid certain foundations for such an exception. The trauma of war, a time when the unthinkable became reality, probably guided a vision of the world in which social robustness should prevail over economic performance. However, experiencing the premise of economic growth, stakeholders quickly confined the charter to the mere exchange of goods, ultimately leading to the creation of the World Trade Organization.[125] This experience can become a warning today: if we set off down the path of sustainable development, we must at least make the effort to place the proposed solutions within a more global context, endeavouring not to overlook their complexity, as inconvenient as this might be. Finally, it comes down to overcome simple adaptation, that is solutions compatible with a system of which we know all the parameters (the myth of sustainable development). We should rather develop adaptability, that is solutions compatible with a system whose parameters cannot be known in full. We shall see that, in questioning the very notion of performance and by opening the field of possibilities through a systemic approach, the suboptimality approach could offer a way out, while still building on some of the achievements of sustainable development.

Let us be optimistic and suppose that sustainable development is the beginning of a deeper movement. Are the efficiencies brought about by the many innovations of sustainable development enough to put us on the right track and accelerate the environmental transition?

3.5 THE ILLUSION OF TRANSITION

The word transition, whether within the context of energy or the environment, evokes the notion of the replacement of one system with another. What is our current status? Renewable energy accounts for an increasing proportion of the energy mix, in an increasing number of countries. This is perhaps the most striking marker of the ongoing transition. But is it truly a transition?

The proportion of renewable energy is certainly increasing. However, the absolute value of fossil fuel consumption continues to rise. On a global level, 89% of the energy that we consumed in 1970 was of fossil origin, with the remaining 11% renewable (hydroelectric, solar, wind, biomass). In 2016, the proportion accounted for by fossil energy dropped to 65%. However, as an absolute value and given that total energy consumption has increased, we have risen from 9,400 TWh of fossil energy in 1970 to 11,000 TWh in 2016, amounting to an increase of 17%.[126] Thus, the new renewable energy only covers a small portion of the *increase* in energy demand.[127] In simpler terms, never before have we generated so much renewable energy; however, neither have we ever before burned so much fossil fuel. The current development therefore cannot be described as a transition, but rather as an addition. How can this be explained?

The Jevons paradox teaches us that frugal technologies do not necessarily lead to an actual saving in resources. An increase in efficiency always leads to an increase in attractiveness. In most cases, this leads to a wider use of the technology or even creates new needs. In the end, this generates an overall increase in resource consumption, despite the efficiency gains per unit of consumption. This is now well established.

Jevons noticed that the introduction of a more efficient energy source, such as coal at the beginning of the 19th century, also made it more attractive. This led to its larger adoption and thus an increase in the global consumption of this energy source.[96] As such, in the United Kingdom, during the Industrial Revolution, the price of lighting was cut by a factor of 3,000, yet the consumption of gas for lamps multiplied by 40,000.[35] As in the example of biofuels cited earlier, this is again a rebound effect: positive feedback on the very mechanism of optimisation.

Such rebound effects can be found in all sectors that relate to energy. Over the past 35 years, refrigerators have increased their energy efficiency by 10%, but the number of refrigerators has increased by 20%. Likewise, during this same period, kerosene consumption per mile decreased by 40% in the aviation sector, but overall kerosene consumption increased by 150%.[128] Finally, when rebound effects are not considered, the display of efficiency gains is self-justifying while leading to an increased exploitation of resources.

Beyond energies, rebound effects can be found in other sectors too. Let us take the case of road traffic. The development of vehicles for hire, intended to help improve traffic flow, has actually contributed to an increase in the number of traffic jams. Between 2010 and 2016, vehicles for hire were even reported to be the main cause of traffic jams in San Francisco.[129] There is actually a variation of the Jevons paradox specific to road infrastructures, known as *Braess's paradox*: when additional roads are added in response to the vast number of cars and the number of traffic jams, the effect is positive in the short term, however, in the medium term these new roads quickly become more attractive and are thus used more frequently. When the Katy Freeway in Houston (Texas) underwent significant expansion, taking it to 26 lanes, traffic jams actually got worse. This is material evidence of the inability of policymakers to take the rebound effect seriously. A monument to Jevons, as it were. In the end, rebound effects are one of the leading causes of the counterproductivity of optimisation.

We can always find evidence of the Jevons paradox in our contemporary societies. We purchase low-energy light bulbs, but multiply the number of light sources, or even leave them switched on because their consumption is low. Likewise, we invest a great deal of money in the insulation of our properties and invent solutions by which to save energy, all while rushing to switch on our personal air conditioners. Consequently, according to the International Energy Agency, four air conditioners are sold every second around the world, and air conditioners and fans currently account for 20% of energy consumption by buildings.

In spite of these factors, some believe that the momentum towards the preservation of the environment is too strong to be slowed. We would merely be at the beginning of a transition, which would soon take effect. This rhetoric can be found reflected in the Kuznets curve: the environmental impact of a country that is escaping poverty would initially increase because its economy would begin polluting. In turn, the increase in the country's GDP would enable this country to limit future environmental degradation, pollution being considered increasingly less acceptable.

Ecosystems would thus be preserved subsequently. To put it in other terms, there is little concern for the environment so long as primary needs remain unfulfilled.[130] The race to environmental well-being should be sufficient to settle the environmental debt incurred during the economic take-off. It would be a form of emergent regulation, governed by negative feedback.

While this bell-shaped dynamic has been verified for some easily detectable and quantifiable chemical pollutants, such as sulphur dioxide or nitrogen dioxide, it cannot be applied across the board to all of the elements found in our environment. It is therefore difficult to calculate this type of progression for more complex variables, such as soil quality or the status of biodiversity. What is more, given the global nature of pollution, making a calculation on a country-by-country basis would have little relevance. More pragmatically, it can be observed that developed countries still consume more energy and that their environmental footprints continue to grow, contrary to the predictions of the Kuznets curve.[131] In fact, on a global level, the richest 10% are responsible for half of the CO_2 emitted on Earth. Conversely, the poorest 50% are only responsible for 10% of emissions.[132] The very notion of the primary need is relative: a contemporary American or Singaporean family will consider that Internet access or air conditioning are now among their primary needs.

The idea that technology and wealth will inevitably lead to a reduction of human impact on the environment is, to say the least, questionable. It nonetheless remains a common rhetoric, which can also justify innovation. The environmental disaster can be a chance for investors and creators of technologies that claim to be "clean".[133]

Let us consider pesticides. Some believe that biotechnological revolutions have lowered phytosanitary treatments.[134] While this may be true in some cases, a prudent approach is required when interpreting available statistics. For example, between 1991 and 2014, the use of pesticides fell by 33% in the United Kingdom (2.9 kg/ha in 2014), it remained stable and elevated in the Netherlands (around 10 kg/ha), and, in contrast, it increased by 138% in China (15 kg/ha in 2014).[135] In France, the agriculture ministry reports that the use of pesticides increased by 25% between 2008 and 2018, in spite of the fact that the "Ecophyto" plan was supposed to reduce their use by 25% by 2020.[136] More concerning, a study by an anti-pesticide association in the United Kingdom indicates that since 1990, the toxicity of pesticides has increased. Neonicotinoid insecticides are thus reported to be 10,000 times more toxic than DDT. It has been estimated that 4% of pesticide sales (i.e. no fewer than 600 million dollars) relate to very toxic substances. These molecules are said to cause 25 million cases of severe intoxication and 220,000 premature deaths each year, according to the WHO.[137] The land area treated with pesticides, the frequency of treatments and the diverse range of pesticides used are said to be increasing.[138] Furthermore, with genetically modified plants producing a pesticide themselves, one would expect to see the demand for containers of pesticides falling in certain countries. In fact, 98% of genetically modified plants grown commercially around the world, purported to bring about a pesticide-free environment, actually produce a natural toxin or are resistant to an artificial pesticide.[35] Thus, pesticides have not been eliminated from the equation. Instead, we have developed alternative means that further justify their existence.

Other genetic strategies, by means of transgenesis or by more classic crossing, could have been prioritised. For example, the multiplication of some viruses can be slowed when the plant makes alterations to its protein-building machinery, a form of natural resistance.[139] In the same way, the movement of viruses within the plant can be delayed if the pores between cells are partially blocked by callose, a natural carbohydrate produced by the plant.[140] These solutions are more respectful towards the environment in the long term, but are also more complex and probably less effective in the short term. The macroeconomic analysis shows that we have chosen the simple solution of short-term performance. In spite of the objective of a pesticide-free world cited by many governments[141] and research institutes,[142] the overall assessment of the ongoing transition remains that of a race to economic performance that fails to first consider environmental constraints.

3.6 TOWARDS AN OPTIMISATION OF EARTH

Freedom of the market and the boundless creativity of humans do not seem to be sufficient in tackling the challenges of the Anthropocene. Some deterrents to the forcing of nature could be, or have already been activated. Ironically, they occasionally come from unexpected sources. A case in point, the American governments of George W. Bush or Donald J. Trump, deniers of the climate crisis, were the most effective defenders of the environmental cause in some instances. For example, by denying the existence of global warming, the United States was well behind China and Russia when it came to building the icebreakers that would soon supply trade via the Arctic shipping route that opened as a result of the melting of the ices.[143] Likewise, the tentative Paris Agreement on climate change, concluded during the COP21, actually became a global geopolitical challenge on the day when Trump's United States decided to withdraw from it.[5]

Some governments have put into place regulations to respond proactively to the challenges of the Anthropocene. While the idea of necessary frugality runs its course, in particular through energy efficiency, the question of optimisation has not been called into question. Let us consider the history of carbon quotas in Europe since their implementation in 2005. Companies initially exaggerated their declared pollution. States therefore distributed too many quotas, and industry hardly paid for their emissions. Second, industry traded quotas with developing countries that did not use their emission credits. Again, industry did not pay for their actual emissions. Now industry is reaching the limits of this system and is accumulating ecological debts. But this time, the governments agree to allocate free quotas so as not to weaken the economy. Carbon quotas, originally intended to limit emissions, formalise a public subsidy to pollution. Let us be optimistic and hope that the European Union's ambitious targets will eventually work.[144]

In addition to the windfall effects mentioned earlier, the carbon quota system is in fact a projection of the framework of the market economy onto nature: placing a price on nature allows us to regulate it, and the use of a common currency enables it to be integrated into the global market. It seems that ideologically it is not possible to respond to the climate challenges outside of the free market. Ironically, while the objective is climate regulation, the proponents of economic regulation have no say. A state tax could have played a more systemic role, for example, by promoting the commons or agroecology. Instead, the neoliberal model was preferred, where even for nature, the invisible hand of the market would do what is right and good for society. The 16-year failure of carbon quotas in Europe shows that the times of finance and economics do not match those of nature.[145] This example also illustrates our path towards global optimisation. From an economic perspective, the laissez-faire approach enables a degree of agility in the short term. However, it is the entire atmosphere, the hydrosphere and the biosphere that now formally fall under the framework of optimisation for human economic needs. This *green economy* is an economy of total optimisation.

In that scenario, nature would thus become a garden in which humans control every element and interaction. The geo-engineering idea illustrates this perfectly. Ironically, from a logistical perspective the greenhouse effect associated with global warming would require us to live in an actual glasshouse with very precise control over temperature, hygrometry and all other parameters of life. We sometimes speak of the *good Anthropocene* as a means by which to describe this planetary garden. The new global power of humans will impose upon us a responsibility to make use of science to invent the future world, as though we were landing on Mars with the objective of creating a biosphere to serve our purposes.

This philosophy is not new. For example, in 1857 Eugène Huzar, said: "As enormous as the Earth may appear to us, it is not infinite, and human endeavour, it is infinite throughout the centuries. But today, with science, this proposition is inverted: it is man who has infinite capacities and the planet that is very finite. Space and time no longer exist through steam and electricity".[146] The rhetoric of the *good Anthropocene* is sometimes presented as a response, perhaps even an opposition to *business as usual*. While integrating environmental constraints, the belief in technology and progress is not really called into question.

If the theory of complex systems teaches us that everything is counter-intuitive, then why not allow ourselves to be caught up in the scenario of the *good Anthropocene*. Let us suppose that humans are capable of developing technologies enabling control on a planetary level. Would they be also capable of foreseeing all of the consequences of their actions with sufficient precision and anticipation to maintain the homeostasis of the environment? Would they have a sufficient amount of time to do so prior to extreme climatic events becoming commonplace? Would they have invented a world with fully renewable and recyclable resources? Human beings can optimise everything, except for time. Herein lies the limit to this scenario.

3.7 A DIGITAL CATALYST?

In a spirit of harmony and optimism, let us suppose that we could overcome the challenges of the Anthropocene, through a combination of authority and pedagogy. To be credible, we must outline a scenario whereby this transformation was not of our own making. It is indeed difficult to believe that we could go against our own nature quite so quickly.

Fundamentally, 21st-century humans are still hunter-gatherers. This is reflected in our own biology. We can still go without food for several days, or even several weeks, as our distant ancestors did during the winter season. However, the end of famines in the contemporary western world renders this ability obsolete. Similarly, when at rest, our eyes settle at a point on the horizon, perhaps a vestige of our status as hunter-gatherers, continually on the lookout for far-off danger. Our cognitive bias towards the short term, which prevents us from envisaging more rational long-term solutions, is probably also a remnant of our hunter-gatherer past, ready to flee at any moment. Quite simply, we are not evolving quickly enough! We must therefore find an external catalyst.

What form could this take? A world that has become too horrible for our evolution to continue at its current pace? The wisdom or activism of new generations? The world guided by a new, enlightened and dictatorial United Nations (UN)? Why not? In an alternate hypothesis, our digital double could play a greater role in constructing the post-Anthropocene human.

We live alongside an increasingly pervasive technosphere. Consider the amount of time you spend in front of a screen, the only genuine question that you ask the receptionist at your hotel ("and what is the password for the Wi-Fi?"), not to mention the robots and artificial intelligence that assist us, the electronic babysitter, or Zora "the robot that gets senior citizens dancing in retirement homes".[147] Could these machines modify our behaviour?

They are already doing so: Consider the automatic switching off of lights by means of motion detection, the battery gauge of electric cars which refills as one slows down and thus encourages economic driving, or social networks that invite us to re-use consumer goods by means of exchange or repurchase. Digital complexity is not merely the product of economic globalisation and technological achievement, on a more fundamental level it is a response to the end of abundance. The temporal simultaneity of the two post-war revolutions, the digital revolution and the environmental revolution, may come as a surprise. These could be the two faces of a singular response to the scarcity of resources.[148]

Although revolutionary, isn't this new position held by technology too superficial to profoundly modify our behaviour? Given the speed at which digital technologies have altered our jobs, our way of travelling and our diet, this is certainly where my bet would lie, at least in the short term. We are entering into an era dominated by algorithms. As such, the artificial intelligence that manages the financial flows at the New York Stock Exchange creates wealth, and also mini market crashes, in the same way as humans would (were they physically capable of acting within a matter of milliseconds). If power is becoming algorithmic, then machines will tell us what to do, and we will be capable of accepting these injunctions without necessarily realising that it is a distortion of our true nature. The post-Anthropocene, the *Algocene*, is perhaps already on our doorstep!

In a world that is facing exhausted resources and degraded ecosystems, apps could prevent us from eating meat more than once a month, stop us checking in for a flight, or place us on lay-off because we have already consumed our carbon quota for the month. Is this science fiction? In China, the social credit system, which uses digital technologies to assess individual reliability, is already imposing such restrictions, albeit for other political reasons.

We could come to regret the degree of control that we have already afforded to machines; however, we cannot but acknowledge that our actions are in complete contradiction to any such irritation. If the digital giants occupy the leading stock market capitalisations, it is indeed we, the consumers, who have placed them there. Within an optimistic scenario, this digital duplication could be transitory, and could perform the role of a catalyst for environmental change by means of the promotion of frugality. At the very least, this scenario stresses that the expected rapid global change will require extraordinary measures.

However, the option of digitally accelerated frugality remains very clumsy. It has been estimated, for example, that a 2-gram microchip requires 32 kg of solid raw material for its manufacture.[149] So what frugality are we talking about exactly? The *Algocene* scenario has at least the merit of highlighting the limits of human capacity to initiate change. Because of its tight link to natural resources, it also illustrates how narrow our future path has become. Here again, we are reaching the planetary limits.

3.8 SCARCE RESOURCES

We are dependent on an ever-increasing number of chemical elements in order for our civilisation to function. A wide range of rare minerals are now found in our modern tools, and in particular computers and other mobile phones, where algorithms would guide us towards frugality. Even if the transition to a cleaner world is an imperfect one, how much time would we have before these non-renewable mineral resources start to become strained?

It is difficult to make such estimations since the players within the mining sector have diverging interests in this respect, with scarcity inflating the prices and abundance maintaining confidence. Nonetheless, here are a few figures for the more pessimistic: indium, which is found in touchscreens and photovoltaic solar panels has a life span of 5 to 10 years; zinc, which is found on roofs and in brake pads has a life span of 20 to 30 years; antimony, which is a flame retardant element has a life span of 15 to 20 years.[150] A common element such as copper, essential in electronics, could become a resource in tension in the next 40 years. The French Alliance for the Coordination of Energy Research estimated in 2015 that it would be necessary to extract as much mineral from the subsoil as humanity has extracted since its origin, to satisfy human needs until 2050.[151]

The estimation of the resources still available is in fact a little more complex than this. In fact, while the Earth's crust still contains numerous elements, we will in the future have to dig deeper, and therefore use more energy, in order to extract them. There will come a day when the return on investment will not be sufficiently high, even though resources are still technically available. A resource is under strain when its cost of extraction becomes too high or when externalities, for example, associated pollution, are no longer acceptable. Through the magic of financial strategies, one can exceed the threshold of that strain in certain cases. For example, the oil extracted from tar sands in Alberta (Canada) almost fully covers the energy required for their extraction today. That industry therefore has no physical added value. Its only added value is financial, with a very high environmental cost. If one looks for concrete signs of an overshoot, the absurdity of tar sands exploitation is a prime example. In other terms, we have picked the low-hanging fruits, and climbing the ladder to pick the others will soon cease being profitable or acceptable.

The Hubbert peak, the curve of which formalises the production of a given element from a stock, enables the identification of a point in time after which demand will begin to surpass production capacity. On a global level, we actually exceeded the peak for conventional oil in 2006. This peak was relatively painless, as it was masked by the emergence of non-conventional oil: in particular shale gas and biofuels. This will not last forever. Indeed, the growth prospects of the shale gas sector are already being called into question.[152] In fact, some predict that the *peak everything* may be reached within the 21st century.[123]

Given the mooted scarcity of resources, their re-use and recycling becomes a major issue. Where do we stand in this regard? Europe produces 12 million tonnes of non-recycled metallic waste each year, and this figure is set to increase by 4% year on year. Of the 60 minerals studied by the United Nations Environment Programme (UNEP), 34 were recycled at a rate of less than 1%.[120] In other words, more than 99% of these elements are simply discarded and diffused into the environment. Incidentally, the United States convened a scientific panel in 2007, the *International Resource Panel*, whose mission is to help nations make sustainable use of their resources. It should also be noted that some metals can be recycled endlessly while others deteriorate in quality and function with each stage of the recycling process. One can acquire an understanding of this deterioration by considering an everyday (granted non-metallic) product such as paper, which can only be recycled 4 to 7 times because the cellulose fibres shorten with each cycle.[6]

While the loss of some of these elements is good news for the planet (*via* negative feedback on our consumption), it also throws a spanner in the works of the good Anthropocene: we will have to build an economy based primarily on our main renewable resource, plant biomass.

The scarcity of non-renewable resources has become a geopolitical challenge. As such, the global economic forum already ranked the depletion of non-renewable resources in fourth place in the list of sources of worldwide concern in 2012 (after the euro crisis, the global economic crisis and the digital revolution[153]). In 2020, the World Economic Forum's (WEF's) global risk report was, for the first time, dominated by the environmental crisis. Companies such as BP, commission reports on the status of resources. In fact, the production of 21st-century energy requires an increasingly complex cocktail of elements, whereas carbon, calcium and iron were sufficient in the 18th century.[154] Here once again, the contradictions of the Anthropocene surface, where some companies that openly advocate growth are commissioning reports on the exhaustion of mineral resources and the associated failure of their economic model.

Some believe that the miniaturisation made possible by nanotechnologies will manage to reduce future needs – a form of geometry-based efficiency. Once again, the rebound effect could well prove this prediction wrong. Furthermore, the majority of current applications of nanotechnologies focus on food, anti-caking or antibacterial products. These new technologies do not limit the use of resources, but rather they promote their dispersion, even prohibiting any possibility of future recycling.[123] Incidentally, this dispersion problem is not unique to nanotechnologies. For instance, modern smartphones comprise 60% to 80% of Mendeleev's periodic table of elements,[155] an indication of the difficulties that lie ahead when attempting to find substitutes for exhausted elements. In 2009, Americans discarded an average of 350,000 mobile phones each day.[156] Given that fewer than 20% of these phones are recycled, the majority of these elements are dispersed into the environment.

Also conscious of this, yet another time bomb, since 2011 the European Commission has been compiling a list of minerals that are essential to the economy of the European Union, and the supply of which is under threat (critical raw materials, CRM). The list is growing with time. Fourteen CRMs were identified in 2011, 20 in 2014, 27 in 2017 and 30 in 2020. This increase can be explained by the fact that some minerals are approaching their Hubbert peak and the cost of extraction is increasing. Aside from the stock and the cost of extraction, regulations or geopolitical conditions can render the extraction or use of some minerals more restrictive. Take cobalt, for example, which is used in batteries: half of its supply comes from the mines of the Democratic Republic of the Congo,[157] a chronically instable country, and its use is subject to increasing restrictions within the European Union on account of its toxicity.[158]

Other seemingly much more abundant elements have also become rare. For example, construction sand, from quarries, rivers and oceans, has an angled shape that allows it to be used in concrete, unlike desert sand, which is round and polished by the wind. It has been estimated that every human on Earth consumes an average of 18 kg of construction sand per day, indirectly![159] It has even become an attractive merchandise for organised crime groups all over the world, since ocean floors, river beds and quarries are being depleted.[160] Between 75% and 90% of the world's beaches are receding due to the exploitation of the oceans' sand and currents that draw the sand from the beaches in order to restore a balanced distribution.[161] Even in the most optimistic scenario, we will not avoid the physical limits of our environment and the scarcity of the elements currently required for our activities.

3.9 AN IMPASSE?

I thus return to my hypothesis: let us suppose that humans in the Anthropocene have changed, or even denatured, perhaps with the assistance of algorithms, and have become conscious of their environment and the multiple implications of their over-consumption. Furthermore, they are taking action in order to align themselves with their new ideology. How much time do we have before this must occur, being both optimistic and potentially totalitarian?

To answer this question, let us focus on the basics: the amount of food resources available on Earth. Our economy, our security and ultimately the rule of law, rely on stable access to food. Let us consider the case of a fertiliser that is now vital in order to maintain crop yields: phosphate. It is thought that we will pass the phosphorus peak in 2030 at the earliest,[162] at a point in time when the human population will be approaching 9 billion individuals.[163] If we were to look only at the rate of phosphate extraction since 2016, the known reserves of phosphate rock would be exhausted by 2040 in the most pessimistic scenario.[164] This might be the greatest threat with regard to human food: without phosphate and without nitrate, global food production would be reduced by half.[165] This is a great incentive to input reduction strategies, especially through agroecology. Viable recycling solutions for phosphorus are emerging,[164, 166] thanks to the phosphate contained in animal bones[167] or to a better knowledge of the plant-fungi symbiosis (mycorrhizae).[168] Despite these promising prospects, the example of phosphorus illustrates the urgency we will have to face.

Beyond fertilisers, the amount and quality of arable land becomes a central issue. Agricultural intensification limits the sustainability of arable land, in particular because of ploughing (which promotes desertification) and chemical pollution (fertilisers and pesticides). Furthermore, not all of the available land can be dedicated to food farming, either because it is not suitable or because it should be preserved to provide other ecosystem services. It is difficult to find a balance between food, demography and the environment, and there are signs that our current trajectory is not sustainable. For instance, world cereal production per inhabitant is declining since 1985.[37]

The issue of food security cannot be reduced solely to the production of food. There are more overweight humans on the Earth today than there are undernourished people. On a global scale, the threat to food security is in fact also driven by waste and distribution problems. Consequently, it is estimated that one-third of all food is discarded or lost each year. The issue of food security thus includes many components, such as local production disparities, distribution and storage capacities, and education. Is there any factual proof as to the threat to our food supply? The 20th century saw remarkable results in the battle against hunger and malnutrition. However, this trend inverted in the run up to 2015: the number of undernourished or malnourished people has risen slightly every year since.[169] Were we merely witnessing a break in progress, or rather a turning point?

While our food may be sufficient in quantity, its quality has altered during the Anthropocene. We are currently living in a kind of chemical soup, where the harmful effects of cocktails consisting of toxic natural elements remobilised by human beings (such as mercury as referenced earlier) and poorly biodegradable synthetic molecules are continually stacking up. Endocrine disruptors contained in packaging or pesticides are regularly under the spotlight.[170] Chronic illnesses are expanding rapidly in the West, as well as in less developed countries. It is difficult to avoid drawing a link between our way of life and the new chemical order of our global environment.

In addition to chronic illnesses, another marker of the physiological effects of this chemical disruption could be found in the measurement of the average intelligence quotient (IQ) of the respective populations. In the western world, IQ increased in the 20th century, however, it has now been on the decline for 30 years. Aside from socio-economic factors, the impact of the new environmental chemistry on our biology could exacerbate this tendency in the future, if nothing is done to prevent it. For example, it has been established that, in the frog, the thyroid of the "mother" is sensitive to environmental chemical factors and that its dysfunction could alter the IQ of the "child".[170] In the same way that CO_2 serves as a marker as to the condition of the climate, the human IQ could become a marker of the harmful effects of our polluted societies. An *Intoxication Quotient*, as it were.

While the production and distribution of food are manageable parameters, the same cannot be said of demography. The action that is by far the most effective in reducing our impact on the environment in the long term would be to limit the number of births. Thus, coldly reduced down to a quantity in tonnes of CO_2, choosing to have one child less, in the present western world, would avoid the production of an additional 58 tonnes of CO_2 per year (when projecting the cumulated the carbon footprint to successive generations). By comparison, a trip around the world by plane produces 3 tonnes of CO_2 and according to the 2015 Paris agreement, all humans on Earth should emit less than 2 tonnes of CO_2 per year for the planet to be within the +1.5°C climate warming. Although the advent of a more "eco-educated" society would inevitably lower this figure of 58 tonnes per year in the future, there are many reasons why the subject of limiting birth rate is considered taboo. Incidentally, school textbooks do not present this option among the possible means by which to tackle the climate crisis.[171]

Some countries have considered this issue. This was certainly the case with China and its single child policy, implemented between 1979 and 2015. Although this form of birth planning has since been abandoned, the birth rate remains low in China, only now for economic reasons. Japan and South Korea have been in this new demographic transition for longer. The ageing of the population is often a cause for concern, even though it could constitute the essential transitional stage for readjusting human needs to planetary limits. There are also some isolated territories, such as the 5 km^2 island of Tikopia in Oceania, where the population is maintained at fewer than 2,000 inhabitants by law.[69] However, these examples constitute exceptions to the rule: a proactive transition towards a reduction of the global human population is not on the agenda.

3.10 DAY ZERO

Considering that human beings cannot live for more than three days without water, the limits of our resources become clear when we consider the issue of drinking water reserves. For instance, according to Foreign Policy, in 1979 Iran had renewable water resources amounting to 135 billion cubic metres, for a population of 34 million inhabitants. Today it is estimated that these stocks amount to 80 billion cubic metres for a population of 80 million inhabitants. Seventeen countries on Earth are currently in a situation of extreme water stress. Alongside Iran, this also includes India and its 1.3 billion inhabitants.[172] This is a country that uses almost all of the water it has at its disposal, that is to say, not a year goes by that they are not on the verge of *day zero*, that is the day when all of the reservoirs and water reserves dry up.

The case of Saudi Arabia is a prime example. The citizens of this country have a direct and indirect water consumption that is two times greater than the global average (263 litres per day per citizen for a global average of 137 litres per day per citizen, according to UNICEF). The causes are well known: an arid climate, a growing population, a non-sustainable lifestyle made possible through access to oil wealth, a cereal farming industry that consumes a great deal of water, right in the middle of the desert. According to Rebecca Keller, of private intelligence company Stratfor, this country is believed to consume "more than four times the volume of water that it renews". Mohammed Al-Ghamdi, a water table specialist at the King Faisal University, predicts that Saudi Arabia will have depleted its water reserves by 2030.[173, 174] Conscious of these challenges, desalination plants have multiplied in number to cover half of the Saudis' water consumption. However, in addition to the fact that these plants are major consumers of fossil fuels, the desalination process also generates brines that are extremely harmful to the local environment.[175]

There is no point in engaging in political science fiction here, but one will nonetheless recall that the initial migration of Syrians peasants to cities within their own country before 2011 was caused in part by the Assad government's catastrophic plan to develop irrigation for the cultivation of cotton (with consequences of which we are now aware). Likewise, the mass migrations of the Hondurans to the United States were in part caused by errors in the management of water, coupled with extreme climatic events such as Hurricane Mitch and the recurring droughts since 2014.[176] To those who would believe that the water dilemma only applies to countries that are severely impacted by drought, even the umbrella country, England, could also suffer from water shortages within 25 years, according to the United Kingdom Environment Agency.[177]

It therefore appears that we will have to confront a fair degree of environmental turmoil over the coming decades. Here I have listed only some of the significant parameters concerning our biological and physico-chemical environment. Yet ecosystems are extremely robust over a very long period of time. In fact, they have existed for billions of years, and they have navigated fluctuations that, while perhaps less rapid, were of greater amplitude (see for instance the chronology at the end of this book). If our drive to optimisation is heading to an impasse, the weak link that will break first is the social system, which has only existed for a few thousand years, within the mild and relatively stable climate of the Holocene.

The Meadows report mentions the weak signals of a society that goes beyond the ecological limits: new professions to exploit resources that are rarer, less accessible, more scattered and of poorer quality (e.g. extracting construction sand from the bottom of oceans, to be desalinated), redirection of labour to compensate for the loss of free ecosystem services (e.g. depollution, protection of biodiversity, pollination by hand), lack of investment to maintain infrastructures and social protection systems because more immediate threats must be addressed, massive debt, increase in conflicts over natural resources, new mass consumption patterns towards basic needs, challenged legitimacy of public institutions.[37] If there is a turning point in this century, it will probably be social before it is environmental.

3.11 AN EXTRATERRESTRIAL HUMANITY

For some, we would have to leave the Earth. While rocket technology is now well mastered, colonising another planet presents far more complex challenges. In particular, space programmes for the colonisation of Mars (or other exoplanets) from NASA and ESA raise the question of our capacity to develop and maintain circular farming.[178] In fact, if the atmosphere of these other planets is incompatible with human life and open field plant cultivation, we will have to take into account all of the material cycles in order to guarantee the viability of human colonisation. To date, we are not able to create such agriculture, condemning any attempt to leave. Ironically, we begin asking ourselves the question of circular agriculture at the very moment we think about leaving the planet. In other words, and to use the expression coined by Bruno Latour, we become terrestrials the moment we want to become extraterrestrials!

Given the looming environmental and social emergency and the anticipated scarcity of resources, and our current inability to develop circular agriculture, the idea of the colonisation of another planet is at best a good topic for science fiction. We are going to have to react, on Earth. If we only have five to ten years before major turmoil begins, then we cannot wait another minute. In a radical scenario, we would have to stop mineral extraction now, cease burning oil, ban aircraft flight and the consumption of beef. This revolutionary scenario is not incongruous: it is more or less in keeping with the recommendations of the IPCC.[179] Meanwhile, we will have to invent substitutes for all of these non-renewable resources, in record time. The development of bags or straws made from renewable materials illustrates the transformation of technologies, mentalities and legislations, but is also perfectly anecdotal with regard to other resources, not to mention the fact that it took until the 2000s before regional and national legislation began tackling this issue.[180]

So what shall we do? Let us have a look at a country that is experiencing rapid economic development: Indonesia. As we have seen earlier, Jakarta suffers from very pronounced air pollution. If you are sufficiently wealthy, you might elect to send your children to a school that tackles the problem head on. This is the case, for example, in the school that actually brands itself as intercultural (*Jakarta Intercultural School*): in the classrooms, fine particle pollution is filtered out using cutting-edge technology. The school is like an oasis of fresh air in the city. This educational and environmental service of course comes at a cost: the admission fee stands at between 20,000 euros (nursery school) and 30,000 euros (college) per year.[181] This school illustrates our extraterrestrial condition with aplomb: it utilises technologies that are adapted for a human colony on another planet, but instead, this is realised on Earth, and for an affluent minority.

We come to the realisation that the predictions from the Meadows report are credible.[182] The Anthropocene constructs a simultaneity of crises that would be difficult to surpass. Even in the event that we are in error when it comes to estimations concerning resources that are actually more readily available than suspected, and even supposing that the stock of these resources is two or three times greater than assumed, we will likely feel the first effects of their scarcity prior to 2050. In fact, we are already experiencing the negative externalities of the climate crisis and the physico-chemical disruption of our environment. However, have we perhaps poorly assessed the transformation ahead? The predominant scenarios of the good Anthropocene and sustainable development might screen out other solutions.

3.12 THE END OF THE RENAISSANCE

Since we find ourselves on the edge of a precipice, let us take a small step back. The solutions mentioned earlier appear poorly calibrated. Between the temporal hiatus, human psychology, the multiple instances of feedback, the need to make use of a digital crutch in order to accelerate change, or the contradictions of sustainable development, something must be wrong in how we negotiate with our planet. In the end, we do not lack solutions. Rather, we suffer from an overflow of counterproductive solutions. Thus, we may have to abandon some of our more deeply rooted core beliefs to identify a credible and inspiring way out.

If current policies are unsatisfactory given the alarming predictions, this is also because the impacts of the Anthropocene bring about changes that extend beyond the short-term timeframe that is so familiar within the decision-makers. For example, an insecticide used to cultivate bananas, chlordecone, is now banned. However, the soils and coastlines of the Caribbean will continue to be polluted with it for another 500 years. Similarly, some nuclear waste will remain radioactive for 100,000 years. There is no political mandate that is adequate to address such lengthy timeframes. Since the technologically minded will object that we will have found a solution to these problems by the end of the century, allow us to cast our eyes back on what has come before.

The question of the Anthropocene transcends the left-right debate of the liberal democracies of the second half of the 20th century. The Anthropocene instead sets the deniers of the environmental crisis against the realists, extraterrestrials against terrestrials.[5] The Anthropocene marks the failure of neoliberalism, the failure of productivist communism, the failure of global modernisation through colonisation or the market economy.[183] While the atomic bomb signalled the end of blind faith in technological progress, after more than a century of intense social struggle against the misdeeds of the First Industrial Revolution, the Anthropocene closes the cycle initiated during the Renaissance, marked by a belief in progress itself.

The Quattrocento is indeed key in comprehending the first steps of the Anthropocene, which unfolded some 500 years ago. In particular, with the invention of monofocal perspective by Brunelleschi,[184] the humans of the Renaissance not only produced more realistic drawings and designs but they also displayed their will to appropriate territories. The Earth became a garden to be landscaped. From an anthropocentric standpoint, the world became "quantifiable". Alberti actually describes perspective as *Commensuratio*.[185] This "commensurable world" gives human beings the power to establish a mathematical link between locality and globality. In this respect, the Renaissance is a revolution. The catalogue of sustainable development solutions seems rather meagre by comparison. Henceforth, the political response to the immense challenges of the Anthropocene should probably assume an entirely different scale, perhaps taking on the form of a new Renaissance.

Is this the route that we are taking? The recommendations of the IPCC illustrate the change in calibre of the decisions to be made: electricity is to account for 55% to 75% of the energy consumption in buildings by the year 2050, the proportion of low CO_2–emission energy in transport should be between 35% and 65% in 2050, compared to 5% in 2020, and from the year 2050 we should be extracting more CO_2 from the atmosphere than we are adding in.[179] Likewise, the National Research Institute for Agriculture, Food and Environment (Institut National de Recherche pour l'Agriculture, l'Alimentation et l'Environnement), a historic player in the green revolution in France, has turned its focus towards agroecology, making ambitious proposals regarding the reduction of pesticide use, co-cultures, agroforestry and new types of livestock.[186] However, no country has implemented an economic, legal and cultural policy that effectively fulfils such recommendations. Therefore, instead of adding up more proposals for tackling the climate crisis, what is actually required is that we completely rethink our model of civilisation.

To make the ecological transition to a new Renaissance could be a useful means by which to make an indelible impression on peoples' minds, and perhaps even provide the initial components critical to the race to optimisation. Nonetheless, one might at this point object that the race to optimisation dates back to a much older era. Archaeological ruins testify to ancient peoples' grandeur and desire for growth. From the Assyrian temples to the Mayan pyramids. The architecture of past civilisations demonstrates a recurring tropism towards the improvement of performance, which has been ongoing for millennia. The development of irrigation and domestication is also an even more ancient marker of this human tropism towards efficiency. So are we in the correct range of response when we speak of a new Renaissance aiming to alter the trajectory of the Anthropocene? Is the required revolution even deeper than we had imagined?

3.13 THE END OF THE NEOLITHIC?

In the early days of the Holocene, some 11,700 years ago, the climate became warmer and hunter-gatherers gradually turned to farming. This was the beginning of domestication and of the anthropogenic shaping of nature, in particular by means of the improvement of plants and animals. Here it should be pointed out that it was most likely the females who turned to farming, since most men likely remained hunters. The initial varietal selection measures were very probably driven by the women. This was the revolution of the Neolithic.

In *Times of crisis*, Michel Serres suggests that we are now coming to the end of this period.[187] He builds his argument on the decline of mass agriculture in the West. For instance, 1% of the population in the United States lives on farms today, a figure that was closer to 40% in 1900. The disappearance of the farmer might be very temporary, in light of the coming rise of the circular bioeconomy, which is intensive in resources and agricultural services. The development of agriculture being the main attribute here, the notion that we are approaching the end of the Neolithic nevertheless appears relevant today, at least for western societies. We have seen how the great acceleration since 1950 marks a socio-ecological bifurcation, and thus signals the entry into a new planetary regime at odds with the last 10,000 years. The Neolithic was born of a natural climatic anomaly. Today, a new climatic anomaly on a large scale, this time of human origin, could draw this epoch to a close, accompanied by the risk that we transition to a much warmer world, and that this may endure for a very long time to come.[42]

The end of the Neolithic could also arise from our new relationship with the planet in the Anthropocene. In fact, through agriculture, sedentarisation and domestication, humans have aspired to dominate nature. Global geo-engineering[59] or the genetic modification of wild species, for example, with a view to eradicating certain mosquitoes, are feasible prospects that are currently being investigated within international bodies.[188] This form of nature, which held the status of an individual and uncontrolled threat during the Palaeolithic period (lightning, the animal predator, poisoning, etc.), acquired controllable resource status in the Neolithic. Ironically, as a product of human control, nature has once again become a threat in the Anthropocene, this time on a global scale.

Instances of environmental feedback, whether in the form of extreme events associated with the climate crisis or the erosion of ecosystemic services on account of the collapse of biodiversity, are all obstacles to future human development. The solutions envisaged by humans, such as geo-engineering or genetic engineering, are all ambivalent technological levers. To cite just one of these risks, mentioned in a report by the French National Biosafety Advisory Council (conseil national consultatif pour la biosécurité, CNCB) in 2017, the possibility of eradicating species in the manner in which, and with the ease with which bio-hackers could achieve it, poses a "genuine question regarding safety and potential proliferation".[189] In this context, it is striking to note the success of Hollywood blockbuster movies in which the plots have gradually evolved from the technological collapse of civilisation due to nuclear causes (from *Doctor Strangelove* to *Terminator*) to a more diffuse geo-biological threat (*The Day After Tomorrow, The Happening*). We may finally draw to close a period during which the main role of science and technology was that of incremental increases in performance.

Thus, what might the new Earthling look like after the Anthropocene? It is difficult to say. Let us consider those who have not yet been affected by the Renaissance, by the Industrial Revolution or even by the Anthropocene. Let us make a short detour to South America, accompanied by Philippe Descola, to consider the Achuar – the Jivaroan Indians living in Ecuador and Peru. For the Achuar, as is the case for many other Indigenous peoples, no distinction is made between humans and non-humans, all have a soul. The Achuar actually talk to animals, plants and spirits. We find in the Achuar a form of animism that is widespread throughout the world, in complete opposition to the idea of culture, a notion exclusive to humanity, and one that would separate the human being from nature. The approach adopted by the Achuar is indeed that of widespread coexistence.[190] It would therefore appear that the fusion of humanity and the Earth is not new. Its roots most likely run deep. It is rather the dualism between human and non-human that is an anomaly in the history of humanity, having first come about with agriculture, followed by industrialisation, and culminating in the Anthropocene, where nature is deemed a useful resource rather than a companion.

Remarkably, human society of the late Anthropocene is suddenly asking itself questions with regard to animal welfare, and even plant welfare. This new position ascribed to nature is also accompanied by a new incarnation of the sacred within the living, with plants, situated knowledge and animism slowly acquiring a new status. For example, to continue unwinding the film reels, movies such as *Avatar* or documentaries met with improbable success just a few years ago, such as *The Secret Life of Trees*, literally preach in favour of a new relationship with nature that aligns itself fairly closely with the animist philosophy.[102] The response to the Anthropocene is therefore not expressed merely in the politico-socio-economic world, but also finds a popular and spiritual expression, the return of the natural sacred, which finds its routes in prehistory – perhaps in the Palaeolithic. What form of relationship with the Earth are we heading towards?

3.14 THE NEW EARTHLING

Let us return to the definition of the Anthropocene. Literally, it means the *new age of humanity*, or strictly speaking, the *new human*. What are we talking about here? In the Anthropocene, humanity has fused with the planet, and the planet has fused with humanity. Our fates are tied.

This proposal is based on our now global impacts, and on our own ontology: we all carry within us small fragments from all four corners of the Earth. Take the already discussed example of phosphate. Two countries hold the main reserves of phosphate, Morocco (38% of the estimated reserves) and China (27% of the estimated reserves).[191] Given that 95% of phosphorus is used as fertiliser, and that phosphorus makes up a considerable portion of our genetic material (it is one of the main elements forming the skeleton of the double helix of DNA), we all have a Chinese and Moroccan DNA, long before any other consideration concerning territorial rights or blood rights comes into play. Conversely, human traces can be found everywhere on Earth, from marine ecosystems to the Siberian tundra. As naturalist Buffon phrased it, "the entire face of the Earth today bears the imprint of human power".[192] The more contemporary example of radionuclides, extracted from the depths of the Earth, is apt: more than 2,000 atmospheric nuclear explosions, on all continents, have spared not a single square metre of the planet.

The Anthropocene invites us to go beyond the "Anthropos", and expand our outlook to include the whole of the biosphere. Ecofeminist Donna Haraway actually prefers to speak of the *Chthulucene*, rather than the Anthropocene. Using the device of imaginary sprawling monsters, called *chtonics*, she predicts the advent of new forms of collaboration between Earth-dwelling species in the future.[193] If our identity as Earthlings goes beyond the status of a mere inhabitant, but rather corresponds to the status of coexisting and entangled entities, then we and the Earth have become hybrid living organisms.

The idea of ascribing a biological character to Earth is not new. For example, in 1776 the naturalist Jean-Baptiste Robinet wrote: "We (humans) and other large animals are vermin of the largest animal that we call the Earth".[194] In the same vein, in 1795 writer Félix Nogaret published "The Earth Is an Animal", analysing geophysical phenomena as though physiological problems.[195]

Adopting a less emotive and more mechanistic approach, in the Gaia hypothesis independent scientist James Lovelock and microbiologist Lynn Margulis presented Earth as a self-regulated system, with its own homeostatic mechanisms, like a living being acting in symbiosis with human activities.[196] More precisely, the Gaia hypothesis is founded on a simplified fictitious model by the name of Daisyworld, in which the surface of the Earth is covered with white or black daisies. Depending on the relative proportion of these flowers, solar radiation is reflected into the atmosphere to a greater or lesser extent. The daisies are meant to regulate the temperature, with the black colour resulting in an increased temperature, and the white colour resulting in cooling by means of the albedo effect. Beyond the specific details of this didactic example, the Gaia hypothesis highlights the feedback caused by the activities of the biosphere, and thus humans. It calls into question the homeostasis of the boundary layer of the Earth, in which life develops.

Finally, the Gaia hypothesis and its various iterations point to the link that is to be established between parasitism and symbiosis. Within parasitism, humans would shape the Earth according to their needs, as in terraformation projects on Mars. In the symbiotic scenario, humans would shape the Earth, but the Earth would also shape humans. Paradoxically, the Anthropocene signals the failure of unilateral human control of the environment. Rather, it points to a future path where it is reciprocity, and its surprises, that will allow us to coexist or not. This is actually a well-known behaviour in ecology and evolution: newly introduced living organisms modify their environment, and in return this modified environment changes the evolutionary trajectory of the organisms. This is known as *niche construction*. In biology, this implies that human beings are not limited by their epidermis but that their bodies also extend into their habitat.[197] What is new in the Anthropocene is that this co-evolution is extended to the entire planetary ecosystem.[198]

If humans no longer live on Earth, but rather in the company of an Earth-Gaia, then something has changed. In other terms, "we are no longer individuals".[199] The Anthropocene is indeed a Copernican revolution, as was Darwin's theory of evolution. Once we have realised the scope of the implications of our activities, we are no longer in the same place in our heads or the same place on this Earth.

The Anthropocene places us in a relativist position: the environment is no longer an autonomous landscape in which humans act and from which they draw. The environment alters the actions and the psychology of humans, who are increasingly intertwined with nature. At the beginning of the history of humanity this evolution was not apparent, with ecosystems providing abundant resources in seemingly unlimited supply. Today, in symbolic fashion, the activists of Extinction Rebellion say, "we are not defending nature, we are nature defending itself".[5] Long before them, in *The Natural Contract*, Michel Serres wrote: "We now depend upon the things that depend upon us".[1] We are now finally on the right level from which to respond to the challenges of the Anthropocene. Essentially, the political decisions to come must answer one core question: how can we coexist with Earth?

3.15 COEXISTENCE

The fundamental anomaly that characterises the Anthropocene is the disconnection between humans and the physical world that surrounds them. This is indeed a paradox. On the one hand, the Anthropocene describes an era in which humans have fused with the planet: human activities have never been so invasive towards the physical world that surrounds them. On the other hand, humans of the developed societies of the 21st century have never been so far removed from nature: they are primarily urban, they spend at least a third of their day in front of a screen and some even believe that milk is a manufactured product, like Coca-Cola. Humans have fused with the planet, but they also have become extraterrestrials. In detaching themselves from nature, humans of the developed societies of the 21st century start to read the signs of a disturbed planet, and respond by progressing more quickly, by means of optimisation. They still promote economic growth and ignore the sustainability of ecosystems.

Our adoption of this mask that conceals reality was a gradual process. Christophe Bonneuil and Jean-Baptiste Fressoz list the main historic steps in the *Anthropocene event*.[35] Here I will mention just a few steps that are key to the western world. For instance, for a long time, Catholics have built majestic cathedrals. In response to these *exuberances*, which Georges Bataille would willingly associate with the dissipation of an over-abundance of wealth,[200] protestants instead seek to invest in earthly human productions. This desire for economy could make one believe that we have drawn closer to the Earth. In fact, paradoxically, it was in considering drawing close to the Earth that humans distanced themselves from it. It was no longer a matter of building for the hereafter, but rather building for the growth of the human population, independent of the ecosystems. In other words, Calvin set the scene for the upcoming industrial revolution.

With the commencement of the exploitation of carbon and mineral mines, geologists quickly acquired the status of natural resource predictors. The advent of expressions such as "estimated resources" marks a first step towards the disconnect from the physical world, since the economies of the respective countries will now be based on estimations, whereas past economies were built on the actual proven resources, physically present within the territories.

A final step in this estrangement from the physical world is catalysed by economic sciences. In order to evaluate the economic power of a country, the Net National Income totals all of the country's revenue and subtracts the capital depreciations. Invented in the 17th century, this type of calculation of national revenue still incorporates a link to the physical world.[201] The fruit of the Great Depression of the 1930s, GDP, only considers revenue and expenditure: it implicitly suggests that the economy functions as a closed circle with consumers and producers, with "free services" provided by nature not taken into account. At the end of the 1940s, the inventors of GDP called into question their only recently created tool, immediately after having observed this limitation. Some proposed the inclusion of the depletion of natural resources in the calculation of GDP, counting mining activity as a negative, for example. This was not the choice that was ultimately made, which actually created the illusion of infinite growth.[35] Reality was virtual long before the computer era.

Finally, it should be noted that our disconnect with the physical world also has a temporal dimension: humanity is developing environmental amnesia. Thus, as *The Guardian* stated in 2019, "if you are younger than 34 years old, you have never experienced a single month with temperatures below the average" established over the course of the past 140 years.[202] The natural environment is deteriorating, but each new generation considers the initial condition of the world to be the (new) norm. The shift is only really discernible when considered across multiple generations.[8]

If the future challenge is to know how to coexist, we will first have to reconnect with the living. Let us therefore pay heed to the Kogi Indians, an Indigenous people of the Columbian mountains, who when visiting Paris in 2018, on a trip outside of their territory, uttered these words: "*How do you learn*?"

The physical sciences, literally *natural sciences*, and the biological sciences, or *life sciences*, can help us. In spite of the multiple disconnects within our society, these fields of study have remained linked to the Earth.[5] A pioneer of bioeconomy, Nicholas Georgescu-Roegen argued in favour of the reconnection of the economy with the real world, by means of physics and biology: "Thermodynamics and biology are the necessary torches to illuminate the economic process . . . thermodynamics because it shows us that natural resources are becoming irrevocably depleted, biology because it reveals to us the true nature of the economic process".[203]

Biology and physics are very much complementary and often describe the same objects from different perspectives. Here we should also note that the early physicians of ancient Greece, such as Anaxagoras, were already raising the idea of such continuity. In the modern world, physics and biology have become distinct disciplines. Yet at the beginning of the 21st century, each took a sideward step, or rather a step towards one another: biology has for a long time remained qualitative, describing beings without necessarily formalising the physiological rules, development or evolution. Today, with the progress made in computer modelling, genetics and imaging, molecular and cellular processes are quantified and formalised by means of mathematical equations, and their validity is tested within a defined parameter range, as is the case with a physical system. Conversely, the physics of inert objects such as soap bubbles or soft matter, finds in biology an unexpected field of play. The key point in this encounter is the consideration of complexity, as defined in the theory of complex systems. This includes the mathematical formalisation of randomness and feedback. Alongside geo-socio-ecological models of the Earth (such as World3), models of the living world are also being developed.

At the crossroads between biology, physics and the Anthropocene, what do we find? The immense power of interactions. Or rather, a knowledge and formalisation of the emergent properties of these interactions. This time, it is not only a question of understanding a living being as an autonomous complex system but also of considering the interactions with its habitat, the entire Earth. Biology and physics thus provide us with the tools to understand the mechanisms of our coexistence with the planet.

What could we extract from the understanding of our new place on Earth? Which of the many variables should take precedence? Which dominant factor could guide our response to the challenges of the Anthropocene?

Our coexistence with the planet in the Anthropocene will have to consider one element above all others: fluctuations. Once again, biology and physics have a lot to contribute on this theme. Everything in biology is dynamic, out of equilibrium, and therefore fluctuating. Physics provides the laws to understand it. The Anthropocene heralds a warmer, more polluted world, with more scarcity and less biodiversity. All these degradations converge towards a common feedback response: a more turbulent world. Thus, after the denial of business as usual, the anger of activists, the negotiation of sustainable development and the depression of the end of modernity, we may have arrived on the road to acceptance. Our world is not going to collapse or go into a form of total control; it is going to fluctuate as never before. We are leaving the era of the *mean* to enter the era of the *standard deviation*. This analysis must lead us to discard some of the solutions mentioned earlier in order to imagine a truly viable path.

Faced with these unpredictable variations, what compass should we adopt? Continuing to optimise, through gains in efficiency and efficacy, could lock us into an increasingly narrower path – a suicidal choice in a fluctuating world. A viable path will therefore necessarily be, on the contrary, a broad path. This is the path of *robustness*, the purpose of which is to maintain the stability of the system despite fluctuations. This is what we will explore in the following pages. Here again, the biological and physical sciences resonate with the Anthropocene to outline and question its trajectory.

4 Suboptimality

We must learn to coexist with the planet in order to survive. How can the living help us to envision a robust trajectory capable of overcoming our tropism towards performance as well as the upcoming fluctuations of the Anthropocene? The concept of suboptimality could enlighten us. In the end, it is a question of deconstructing the myth of a necessarily positive performance in order to better understand the inner workings of earthlings' robustness.

4.1 IS OPTIMISATION AN ILLUSION?

As we have seen in the first part of this book, increments of efficacy and efficiency are consubstantial with the major acceleration that is characteristic of the Anthropocene. I will not return to this point, which feeds into my initial line of questioning. Instead, I would like to make, perhaps, a more direct assessment of optimisation, on the basis of what it purports to be: Have efficacy and efficiency gains really favoured human development in past decades? Is optimisation a fantasy?

First of all, the performance increments of the successive industrial revolutions have not benefitted everyone equally. Colonisation, a marker of the Anthropocene, has fuelled a certain form of economic growth for western societies, in which per capita income has increased by more than 200% in the 19th century. However, per capita income in Asia only increased by 1% during the same period. Performance increments are asymmetrical: peripheral territories are exploited in order to enrich the colonising country. In fact, the impoverishment of colonised countries, as a result of the exploitation of resources, has been accelerated by a distortion of the commercial system, these countries focusing their production lines on producing for export, in spite of the fact that famines were (and in some cases still are) raging. For example, rice exports doubled during the final third of the 19th century in India, in spite of the country having suffered terrible famines in 1873 and 1898.[204] One would have to be a major cynic to see any form of global optimisation whatsoever in the history of the economic take-off of the western world.

Let us imagine that the western nations had exploited these resources in a fair manner. Improvements to processes would no longer be marred by the colonial past. In this case, what is the net balance of the economic take-off of the western world? The English Industrial Revolution enabled the United Kingdom to accumulate 20% more wealth than France in the 19th century.[35] While this certainly constitutes a significant relative increase in wealth, the balance sheet is poor in light of all of the human and environmental consequences cited earlier. For instance, France, then very rural, produced four times less CO_2 than the United Kingdom in the same period of time. If we measure performance only by generated wealth then this balance sheet is, to say the least, open to criticism, and it paints a very laborious picture of industrial optimisation.

DOI: 10.1201/9781003510918-4

Some will say that this is the old industry of the 19th century – we can do better in other sectors now. Let us then consider the agricultural sector. Mechanisation and chemical fertilisers have quite literally been the fuels of the green revolution, marked by incredibly rapid improvements in crop yields.[45] However, when the caloric cost of providing oil for these machines and these inputs is incorporated into the final balance sheet, agriculture became a loss-making activity in France from 1970 onwards: it uses up more calories than it produces! In observing process optimisations and taking the final balance sheet into account, actual increases in efficiency are thus relative, even negative: in 1929, French agriculture consumed 1 calorie for every 3 that it produced.[35] In fact, agriculture was counterproductive long before the modern green revolution. According to the UNEP, humans have turned 2 billion hectares of formerly productive arable land into wasteland and deserts over the past thousand years,[205] which is more than all of today's cultivated land.[37]

If there is any one sector in which progress is undeniable, it is medicine. We have seen the eradication of smallpox, the rapid containment of certain epidemics (e.g. Ebola in western Africa), which in times gone by would have resulted in millions of deaths, and the development of new and particularly encouraging transplant and surgical protocols.[54] Do we have here a concrete example of successful optimisation? Health should not be limited to medicine alone. The WHO actually defines it as "a state of complete physical, mental and social well-being and not merely the absence of disease or infirmity". Today, chronic illnesses have become the leading cause of mortality in the world.[170] In this context, it has been estimated that air pollution prematurely kills 8.8 million people every year.[206] The fact that 92% of the global population breathes excessively polluted air led the director of the WHO to state, "air pollution is the new tobacco".[207] This form of pollution indeed kills 16 times more people than violence, 19 times more than malaria and 45 times more than alcohol.[208, 209] Likewise, the healthy lifespan index was on the rise until 2007, and since then has been in decline.[35, 209] Childhood cancer occurrence is well reported in the media: over the past 30 years, its number has increased by 1% to 2% each year in Europe. This can be partly explained by better detection. Yet these impressive figures might also reflect a degraded environment for the child and the pregnant mother.[210]

With the emergence of the environmental question becoming inevitable, it is also increasingly being incorporated into assessments of public health. It is likely that the health sector will integrate more distant objects in the future, such as the quality of water, air and soil, or even the preservation of native habitats for non-humans to counter future pandemics. Conversely, we could learn from non-humans for our own health, for example, by drawing inspiration from the animal pharmacopoeia. For example, ethno-veterinary studies reveal the ability of elephants in Laos to provide their own healthcare by drawing remedies from the forest.[211] In other words, we are currently transitioning from the perspective of human health to that of planetary health, another marker of human-planet fusion in the Anthropocene. With this perspective, the rapid improvements in medicine no longer appear to compensate for the deteriorations in health caused by the degradation of our environment.

Let us finish by considering the information technology sector, an industry that is supposed to lead us to zero-paper status and bring about a reduction in business travel thanks to video conferencing. Between 2000 and 2010, the consumption of paper only dropped by 1.3%, and, as with tourist travel, business travel exploded.[123] The expected frugality brought about by the digital revolution is part of the same mirage. For instance, Google is using more than 16 billion litres of water per year to cool down its data centres in the United States.[212] Here the semantics are quite fallacious: clouds should instead be called data heaters. While 58% of Internet traffic is already accounted for by the broadcasting of videos via streaming services, the anticipated take-off of *cloud gaming* could heighten the energy demand even further in the years to come.[213] In fact, it has been estimated that the sector of information and communication technology could account for 20% of global electricity consumption by 2030.[214]

It is thus becoming clear that the increments of technological progress are not always guided solely by the supposed qualities or benefits of these technologies. This can indeed be verified by considering the history of the Industrial Revolution. For instance, the textiles' industry switched from hydropower to coal during the course of the 19th century, and this in spite of the fact that only 5% of the capacities of British rivers had been exploited and hydropower was likely more economical. One of the main reasons behind this switch from water to coal relates to the dispersion of the rural textile factories that were powered by hydroelectric power. In a remote village, loyalty must be established within the workforce in order to sustain it. Industry that is dependent on hydroelectric power is therefore more vulnerable in the event of social conflict. In contrast, coal-dependent industry is urban, centralised and therefore more protected in the event of social conflict because workers are captive.[215] In the same way, the transition from coal to oil was not caused solely by the intrinsic qualities of oil. The social component played a significant role in reaching this turning point, the transport and extraction of oil requiring a level of technicality that distanced it from the unions.[216] Technological advances are not always motivated by the rationality of so-called progress, but rather by social or political considerations. Optimisation is financial before it is technological.

4.2 COUNTERPRODUCTIVITY

Technical progress has always raised concerns and fears. As sociologist Ulrich Beck wrote in *Risk Society*: "The social production of wealth is systematically correlated with the social production of risks".[217]

Many thinkers in the field of political ecology have voiced criticism of the race to progress. The case of Ivan Illich serves as a prime example, as it extends this criticism to organisations and the associated counterintuitive consequences. Initially earmarked as a diplomat to the Vatican, he ultimately settled in the United States where he taught at the Catholic University of Puerto Rico. His experience of the institutions in Rome and this switch to the field of education enabled him to lay the foundations for his critique of organisations. Not only did he highlight the similarities between school and church, he also observed that schooling does not reduce, but rather promotes social inequality. One could talk of hyper-optimisation and the associated negative effects. Although the initial goal is always stated (in the case of the school, that of reducing social inequality), an optimised organisation impedes this objective when it becomes too efficient. In other words, beyond a certain level of optimisation, performance becomes negative. Ivan Illich ties in with Jevons and his paradox in this regard.

Needless to say, this criticism is not limited exclusively to schools: "Because, above a certain threshold, the tool transforms from a servant into a despot. Above a certain threshold, society becomes a school, a hospital, a prison. And so begins the great imprisonment".[218] With Jean-Pierre Dupuy, Ivan Illich formulated the concept of *counterproductivity* to show how the Industrial Revolution was, first and foremost, the history of a destruction of social links, making way for a civilisation of technique and engineering.[219] For example, counterproductivity explains how, in building cars that can travel at ever-increasing speeds, we waste more and more time in traffic jams. Similarly, companies create competitive environments for their employees in which burn-outs end up being detrimental to their function. The digital economy is no different. Thus, the possibility of same-day delivery offered by Amazon causes a number of fatal accidents among delivery drivers or other road users, all to satisfy the whims of consumers.[220] Counterproductivity is also relevant to the most essential good of societies: food. In 2012, Olivier de Schutter, United Nations special rapporteur on the right to food, noted that, "our food systems are making people sick". In this case, the performance of agro-business has also generated a global and multifaceted pandemic: diabetes, obesity, reduced nutritional value of food products, pesticide and nitrate pollution, agribashing and suicide among farmers.[124]

The idea that optimisation can be counterproductive is, in fact, even more widely recognised. Thus, according to Goodhart's Law, "when a measure becomes a target, it ceases to be a good measure".[221] In other words, the measurement of a piece of data subject to a form of competition is rarely reliable, as it is self-justifying. Goodhart's Law applies in all sectors, and is founded on the race to optimisation. For example, the New York Police Department admitted a manipulation of crime statistics to give the impression of an improvement in security.[222] Likewise, it is well established that education objectives break down when the grading of school pupils becomes a goal in itself.[223] As we have seen earlier, whether it is sport competition and its systemic doping, or rating agencies in finance, the objective of performance ends up overriding other objectives better related to people's lives. For the sake of completeness, this counterproductivity of measurement is also present in my own professional field, that of scientific research. This field is increasingly reduced to journal impact factors, the number of article citations or the *H-index*, intended as a means by which to review the performance of researchers.[221] This race to the top has mainly stimulated experimental shortcuts and fraud. The term *over-optimisation* has actually been deployed here.[224] Ultimately, optimisation ends up driving the competition, and this dynamic is self-sustaining. In other words, once we reach a certain stage, optimisation slips from our grasp and becomes a self-justifying mirage.

The conclusion here is not that increases in efficiency have been negligible, or even losses have occurred. Let us simply say that optimisation in the Anthropocene has rarely been equitable. It has occasionally been beneficial, although not always. It has generated global pollution and it has created the conditions for a collapse of biodiversity, threatening the viability of humanity on Earth. The displayed optimisation is therefore rather a fantasm. On the one hand, the objectives to be reached are not achieved, and are even impossible to define in a world that becomes increasingly unpredictable (no efficacy gains). On the other hand, competitivity and productivity gains are most often counterproductive when considering the broader and longer-term impacts (no efficiency gain). Because it is self-justifying, optimisation quickly turns into over-optimisation, and this over-optimisation weakens the system. The discourse of history textbooks painting the mounting progression of knowledge-based societies would thus require closer and more nuanced analysis.[225]

However, the fact remains that we are still living under the ideology of an optimisation that is for the most part positive, based on efficacy and efficiency increments. Looking beyond socio-economic sectors, this optimalist ideology can also be transposed onto education, in particular through the idealisation of life.

4.3 LIFE AS A FANTASY

Biological systems are often presented as little miracles of organisation and industry. Do we not speak of "metabolic factories" to describe plant biochemistry, or use the term "lines of defence" in describing the immune system?

Let us consider the case of bacterial movement. Some species, such as salmonella, have a flagellum, that is a kind of flexible tail positioned on one side of the cell. The rotation of this flagellum propels the bacterium at a remarkable speed, travelling up to 60 times its own length every second.[226–228] If we were to have such flagella on our feet, in the same proportions, we would be capable of swimming at a speed of 60 × 1.70 m = 102 m/s, or 367 km/h (228 miles/h)! The key to this motion is a molecular machine in which a structure comparable to a motor can be identified, with a proteinaceous rotor that turns at a speed of 10,000 rpm. While these figures may be impressive, to measure the performance of a bacterium on the basis of its speed, or the ingenuity of its motor, would be reductive to say the least. This would be to overlook all of the other aspects of its life, somewhat akin to reviewing the performance of a writer on the basis of his typing speed.

Viewing the world as a machine correlates with our detachment from reality. For instance, in our conversations, we are in "stand-by mode" or in "seduction mode", as if we had access to our own switches. In a perhaps less intuitive way, this is also the standpoint of creationism. This theory indeed postulates that God is the creator of all species on Earth, and therefore stands in direct opposition to the theory of natural selection of species articulated by Charles Darwin.[229] It should be noted that creationism is making a comeback today, most notably under the force of American evangelists. In particular, the concept of *intelligent design* has become part of a pseudo-scientific rhetoric intended to sophisticate that theory: the creator would have produced perfect machines, unattainable through natural selection. The case of the flagellum on bacteria is incidentally often cited in an effort to explain how evolution could have never resulted in an object of such complexity. This argument is easy to counter, since ecosystems are considerably more complex than the flagella of bacteria, and we now have evidence that they are evolving in real time in accordance with Darwin's law, not to mention the fact that the premise of a creator God is tautological, and therefore not explanatory. However, the very existence of creationism, in an era in which scientific fact has formally confined it to the rank of a metaphorical fable, is a marker of the idealisation of life within our societies.

On a more fundamental level, technical or creationist perspectives of life are reassuring in their implicit stability. Yet if you speak to biologists, they will invariably utter the word "dynamic", regardless of your question. If life is ideal, then this is a fluctuating, instable and impermanent ideal. In short, life is continuously imperfect.

4.4 LIFE IS MOTION

Stability (compatible with an ideal of perfection) and dynamics (implying a continual imperfection) have driven the discussions of biologists over the course of past centuries. The realms of living beings were classified according to their phylum, class, order, family, genus and species by Linné and his successors. Despite the implicit stability, this index is useful as a means of identification and in that it provides points of comparison. Progress in physiology has not fundamentally altered this point of view, even though living organisms were already considered objects with internal dynamics.

A few decades before Darwin, Lamarck, a promoter of the theory of transformism, made the revolutionary proposal that species are caught within a temporal flow: we are but observers of a world in perpetual change.[230] Like Darwin, Lamarck implies that a given living organism is more or less adapted to its current living conditions, and that it will not retain its attributes over the generations. It is thus continuously imperfect. In comparison with Lamarck, Darwin goes even further in adding that this flow of evolution is blind, without a purpose, responding only to the laws of chance. Natural selection is all that is required to provide a direction for this trajectory, which is determined only after the fact.[229]

The idea of a selection of the most suitable could lead one to think that living beings optimise themselves in the course of evolution. We find here a very biased reading of Darwin's essay, and finally very much influenced by the industrial ideology of the 19th century and its aftermath in the contemporary neoliberal world. *On the Origin of Species* is the bedside book of competition fanatics, who clearly have not understood the theory of evolution. Indeed, Darwin does not say that only the best individuals are selected, but rather that organisms with satisfactory features can make it through. Darwin thus implies that it is not the most adapted individuals who survive, but rather the most *adaptable* ones. This nuance changes everything, since it is not a question of being the most competitive, but on the contrary of keeping some slack! Similarly, a superficial reading of the theory of natural selection could lead one to believe that it channels a form of homogeneity, the best being the only ones selected. Darwin says exactly the opposite: for natural selection to operate, a strong heterogeneity in the population is required. Between slack and heterogeneity, natural selection therefore requires a large degree of non-efficiency.

Darwin also buries the notion of efficacy in biology by proposing that evolution has no goal. If there is no goal to reach, there is no efficacy, by definition. This may seem surprising for humans developing intentions, which are emergent properties of our interactions and internal physiology. Humans have conscious goals. The lack of purpose in evolutionary biology is actually easier to grasp if we take another standpoint: it would be hard to imagine that our haemoglobin or red blood cells would have a conscious purpose to evolve into haemoglobin 2.0 or multicoloured cells. Again, evolution is basically blind. The idea of a preconceived evolutionary trajectory is a biased anthropocentric conception that would have all living things – plants, fungi, insects etc. – intend to evolve over many generations towards a previously identified destination. In biology, an unforeseen destination is reached. Evolution only builds on the path taken.

Last, the agility of natural selection should not be romanticised too. On the contrary, the evolution of the species has mostly culminated in numerous impasses. For example, in the lineage of the horse, the genus Hipparion colonised a large proportion of the planet, yet it became extinct during the Pliocene, without descent. Evolution is above all the story of a multitude of decimations. The idealisation of the life is thus a twofold delusion: not only is life continuously imperfect, what's more, its method of evolution is not reliable!

4.5 IMPERFECTION ON EVERY LEVEL

Perhaps a form of optimisation does exist in biology, just on a smaller scale. To what extent does the idea of continuous imperfection still apply? In biology, are there not molecules with perfectly predictable behaviour?

With advances in microscopy, it is now possible to observe the trajectory of individual molecules within cells, or to watch the conformational changes of proteins in real time. What are scientists observing? Molecules following incoherent trajectories, proteins that fold and unfold without apparent reason, random collective behaviour, etc. In short, it appears that life is just as dynamic and highly random on a microscopic level. This applies to such a degree that a new field of biology has emerged, *stochastic biology*.[231] In addition to the microscopic level, it now covers other sub-fields of biology such as genetics,[232] developmental biology[233] or population biology.[234] Following the years of "central dogma" and "master genes", suddenly heterogeneity, randomness and variability are the hot topics in biology and appear to provide explanations that are much more satisfactory than the determinist vision instigated during the time of the discovery of the structure of DNA and the genetic code.

At this point, some might argue that life has nonetheless proven its robustness over the course of 4 billion years. So why would one highlight its imperfect nature? So far, biology has mainly been used to serve the ideology of positive performance. This is particularly obvious in biomimicry. This includes the famous Velcro, from the French "velour-crochet" meaning "velvet-hook" and inspired by the excrescences found on the fruit of the burdock, intended to facilitate their dispersion by animals.[235] It also extends to new research projects around the molecular mechanisms of sunlight capture, which could trigger new ways to respond to the energy and climate crisis.[236] While biological examples offer an inexhaustible source of creativity for humans, the fact still remains that the occasional and specific piece of pretty engineering does not make for a robust overall system. Our obsession for performance in living systems says a lot about us, but nothing about life.

Let us consider the example of photosynthesis. This is a process that is required for practically all life on Earth, and upon which humans are entirely dependent for their survival. Is this a little gem of efficiency and industry? In theory, photosynthesis only converts 2% of the incident energy from the sun into metabolites. Furthermore, under real conditions this rate is actually between 0.3% and 0.85%![45] By way of comparison, this is much lower than the yield of a modern solar panel (10% to 30%[237]).

We are thus faced with a paradox: we live within a society that rewards performance, and all the while performance is becoming a major cause of burnouts. Likewise, the optimisation of processes is a cardinal value of our modern society, and yet assessments are mixed, sometimes negative, when considering the initial objectives, and even catastrophic, if one considers the long-term environmental impact. If we are to search for a solution to this paradox within the living world, the response is disconcerting. Living organisms are often cited as the models for optimisation to be followed, yet they are by no means as efficient as one might think. So what does life do if it is failing to optimise in line with evolution? Could it perhaps offer an alternative path?

4.6 INCORPORATING MULTIPLE CONSTRAINTS

The evolution of living organisms is the fruit of multiple constraints. Within a body, any acquired benefit to a given organ during evolution can reduce performance for another organ, even compelling it to change function. Let us consider an example. From an engineering perspective, birds' wings could have, in theory, gradually evolved so as to fly better. Instead, the current consensus in evolutionary biology says that the first wings and feathers appeared as excrescences facilitating thermoregulation, or as a means of attracting sexual partners, functions that later deviated to enable the function of flight.[238] Living organisms are complex systems. Because they host a plethora of interactions, a local modification can have global and often counterintuitive effects. In this regard, with biology and its million years of evolution, we have an immense archive with which to analyse the wealth of these modifications and their consequences.

Would not wings, and their specific features, be too anecdotal? Let us consider an example shared by all multicellular living organisms, the boundary. This notion was coined by biologists by means of a geographic analogy. Initially, it was defined from a morphological perspective. For instance, the thorax of an insect is separated by the abdomen by means of a straight boundary separating the segments of the body. In the same way, in a flower, the area housing the stamens (the male organs of the flower) is separated from that of the pistil (containing the ovules). With advances in molecular biology, these boundaries then acquired a chemical component, certain genes and molecules being more abundant on one side than the other. We even speak of cadastral genes, which keep these areas separate, like border guards as it were. Thus, when the *Superman* gene is absent, the stamen area invades the pistil area, resulting in the production of highly virile flowers that are almost completely filled with stamens.

However, this carefully arranged cartography masks a rich and much more refreshing complexity. As such, biological boundaries are not just sterile barriers, but rather hubs of interaction and production. Let us consider morphogens. These molecules are responsible for identity, and therefore indirectly for the shapes of the organs produced on either side of the boundary. More precisely, a morphogen is synthesised at one site and is diffused into the tissue, rather like a drop of ink in a glass of water. The concentration gradient determines the cellular identities. The cells that are closest to the source of the morphogen have different identities to those that are further away. This is how the *Bicoid* morphogen in the embryo of the fruit fly determines the differentiation of the head and the tail of the insect. We thus refer to "positional information".[239] During their development, organisms contain several sources of morphogens, and the interactions between these gradients of diffusing morphogens generate boundary zones where the morphogens meet. The story could stop there: development would emerge from the combination of these morphogens. In fact, these areas of contact between morphogens are often the site of the differentiation of new centres for the production of secondary morphogens.[240] In biology, the boundary is not merely a barrier, but it is also a gateway to a richer complexity.

This idea of the generator boundary is more simply and more macroscopically illustrated in plants. The branches of trees emerge at the knots, at the axils of leaves. This particular area is a boundary between the stem and the leaf. The buds are also located at this boundary site, and they contain cells capable of generating new organs. Thus, the boundary is generating new tissues and organs in plants, too. It is very easy to recognise these buds: on a tree branch, lift a leaf and you will find at its axil a small dormant bud, the fruit of a boundary. These buds also explain the developmental plasticity of plants: an isolated tree and a tree in a forest will not have the same architecture, since its buds will be dormant to varying degrees. Boundaries, in the form of buds, are therefore areas of plant regeneration and adaptability.

The boundary in biology plays various roles: separation, differentiation, regeneration and adaptation. There are no doubt other functions yet to be discovered. This therefore represents the main distinction from optimisation from a human perspective: biology confronts complexity in order to survive, and is therefore continually integrating multiple constraints.

4.7 INCORPORATING THE LONG TERM

If evolution acts like as an ambivalent mechanism for channelling and diverting living innovations, this leads to the natural conclusion that living beings modify their performance by incorporating a very long-term perspective. If a biological innovation proves itself over a period of just 10,000 years, it will probably be forgotten. Thus, one can very well imagine that innovations have emerged since the last ice age among the organisms currently alive on Earth; however, the majority of these innovations have been lost.

Likewise, innovations disappear and reappear in different forms. Hence, certain studies suggest that the placenta in vertebrates was invented more than 100 times independently. For example, shark embryos develop through placentation, similar to that in mammals, *via* a structure that resembles a placenta, but which was invented in a completely independent manner. Reptiles also have several types of independently invented placentations.[241] Here we observe once again the Lamarckian and Darwinian perspectives of life: living organisms are not stable, but are rather caught up in an evolutionary flow. In other words, we all carry within us the biological memory of our ancestors, and we are continuously subjecting this to the test of our environment.

With progress in the field of genetics, we now have molecular evidence of this mechanism. Thus, classic genetics teaches us that genes are mixed and transmitted from one generation to the next. The study of epigenetics adds another level of complexity: environmental conditions can also superficially modify genes, and these modifications can be transmitted from one generation to the next. More specifically, the gene always has the same DNA sequence, but small chemical modifications activate or disable these genes.

You might argue that this is not a revolutionary idea: it is well established that genes are induced or suppressed by environmental conditions. For example, the insulin gene will be induced if the sugar level in the blood increases. This enables the adipose cells to store this surplus sugar in the form of fats, and thus maintain the blood sugar at its base. Where epigenetics is truly revolutionary lies in the trans-generational permanence of these modifications.[242, 243] In Drosophila, measurements have shown that such modifications can be maintained for up to at least 50 successive generations![244] Such intergenerational transmission is also observed in plants and fungi, but this seems to be less the case in vertebrate animals where a reset seems to eliminate many of these intergenerational changes.[245]

Considering the examples of wing evolution and epigenetics, it is apparent that biology incorporates the long term. In contrast, humans have difficulty planning beyond their own lifespan, beyond one generation or even beyond a political term. In other words, living organisms are capable of incorporating the long term, through the memory of past adaptations. In the Anthropocene, we are becoming aware of the acceleration that we are causing, and the difficulty that we experience in addressing this new tempo. There is therefore perhaps something else to be learned from life, and from its capacity to integrate multiple and unpredictable constraints over such a long period of time.

Before exploring its foundations, I would like to share a particular moment that prompted me to develop a critique of performance and optimisation in order to propose an alternative path that incorporates biological knowledge.

4.8 A RESPONSE TO SMARTNESS

During the *Anthropocene II: The Technosphere Issue* campus organised by the House of World Cultures and the Max Planck Institute for the History of Science in Berlin in 2016, computer scientist Stéphane Grumbach and I were invited to discuss algorithmic intermediation, exploring how the digital revolution can find its roots in the Anthropocene. In particular, we were to address the complexification of interactions within the digital world, as a strategy of frugality in the face of the exhaustion of resources, as well as the genetic complexification of living beings when faced with the scarcity of resources in their environment.[148] For this seminar, we teamed up with colleagues who specialised in *smartness*.[246]

The smartness of *smart cities* supposes a form of process optimisation acquired through digital development. Thus, a smart city will supply an adequate quantity of energy, in the right place and at the right time, in order to serve the needs of a consumer, without any losses in the distribution network. For example, the motion detector lights already mentioned are a very much embryonic form of frugality driven by the idea of smartness.

The most evolved forms make use of Big Data and artificial intelligence. Thus, by compiling millions of pieces of data, a "smart" urban system can modulate road traffic by modifying the sequence of traffic lights, by opening up additional routes and/or by imposing a lower speed limit in certain areas. In medicine, smartness assumes the form of the personalised doctor: rather than prescribing the same cocktail of drugs for certain illnesses, artificial intelligence reads the symptoms and factors (weight, age, history etc.) applicable to the patient in question and suggests a personalised course of treatment. Smartness is therefore a very advanced form of optimisation.

The results speak for themselves. Google has developed methods that detect metastases in the breast with 92.4% accuracy.[247] Certain cases of stenosis are now better predicted and cured using artificial intelligence than by human doctors.[248] Smartness thus offers a tangible benefit in the short term.

Incidentally, the word *smartness* does not leave any room for criticism: What argument could one have against being smarter? Moreover, in desiring to respond to current environmental challenges by means of frugality, smartness also wraps itself in strong moral virtue. Nevertheless, the leading advocates of smartness happen also to be the producers of "smart" technologies. For instance, IBM's slogan is *Think smart.* Or how to create a conflict of interest from within, by appropriating the notion of intelligence. More concerning is that personalisation goes hand in hand with the total surveillance of each of our actions and movements, and of our innermost being. Between the *likes* and other stars pinned to our profiles, the dogma of performance improvement in the digital world automatically implies the generalisation of mutual evaluation. Because it can be the strong arm of a system of control, smartness might carry the seed of a future dictatorship. As novelist Marc Dugain puts it: "The Stasi dreamt of it, the Internet turned it into reality".[249] It can also be noted that this new form of diffuse authoritarianism is emerging at a time when the model of liberal democracy is flagging. The technological attractor invites us to embark upon the path of control, and we are quickly abandoning our now obsolete personal freedoms. The best proof of this worrying dynamic is becoming manifest in the forms of resistance it generates. For instance, the Committee for the Liquidation and Subversion of Computers (Comité liquidant ou détournant les ordinateurs, or CLODO) appeared in 1980.[250]

During the course of our seminar in Berlin, the counterproductive effects of smartness were considered. In order to illustrate this critique in a light-hearted manner, we proposed that the participants create a "disaster app", that is, an application that, beneath the cover of the best possible intentions to respond to the challenges of the Anthropocene, could actually result in an even greater catastrophe. The film *Gattaca* inspired some of the groups, in a world where imperfection and randomness would no longer be accepted. In spite of this framework being provided for the exercise, some participants created apps whereby the counterproductive aspects were not immediately visible, indicating how difficult it is to predict and engage in a destructive downward spiral. For instance, one app predicted the imminence of an earthquake, another involved virtual canaries that, like real canaries used in mines in bygone times, would signal impending environmental danger in the towns through their birdsong.

One group stood out. When their app was to be presented to the other participants, all of the members of the group remained completely silent. And they persisted with this silence. For quite some period of time. Finally, questions from the other groups broke the silence: "Who is speaking for your group, have you prepared anything?" To this question, and those that followed, the response from the members of the group was invariably, "I would prefer not to", as coined by Bartleby. In the eponymous novella by Herman Melville,[251] Bartleby is a hard-working and conscientious clerk, but gradually refuses to perform the tasks for which he is responsible, until he finally refuses his own dismissal. Escape becomes an active strategy in the battle against authority. In our seminar in Berlin, the dramatised refusal to complete the exercise provided an opportunity to consolidate our critique of smartness. The exchanges between the group and the participants during the Bartleby episode were recorded using a mobile phone, kept clearly visible to everyone, and they were played back in full when subsequently debriefing. Within the context of the seminar, the recording of the initially incredulous, then subsequently amused or annoyed reactions of the audience revealed the violence of the request (to use your brain to construct a disaster) and the necessary resistance that it implies. Opposition to an increasingly more invasive technosphere was also expressed.

However, while this resistance is redemptive, it does not necessarily open a working trajectory. While diagnosis is required for recovery, it is certainly not sufficient. Advocates of smartness preach the benefits of control, personalisation and optimisation on every level. However, these potential benefits are limited to an, often, fantastical, short timeframe. In the medium and long terms, smartness could very well be counterproductive.

Thus, are there any long-term benefits associated with the *absence* of smartness? Do we have any evidence that a lack of optimisation could be beneficial? Here we find a new variant of the resistance to industrialism and English positivism initiated in particular by historian Jules Michelet.[252] He was already clearly pointing towards feedback between humans and Earth, served as a precursor to the Gaia hypothesis and condemned the effect of technology on minds.[196] His acute perception of counterproductivity is still relevant in the age of digital technology and smartness, especially when he writes: "The mechanical genius that has simplified and enlarged modern life in the material order is hardly applied to the mind, without weakening and irritating it. On all sides I see machines coming to our rescue to dispense with study and reflection and make you believe that you know".[253] A scathing response to IBM's slogan.

Recent advances in the field of biology teach us that the strategy of optimisation on every level is not what biological evolution has selected over the course of billions of years. Perhaps what we have here is an escape route, and another means of coexisting on Earth in the Anthropocene.

4.9 SUBOPTIMALITY: A DEFINITION

Over the course of millions of years, life provides evidence of adaptability to environmental fluctuations over the very long time. We thus arrive at the central question posed by this essay: What solutions have living organisms developed to sustain themselves and even develop within a fluctuating environment?

One answer lies in the definition of suboptimality in biology. Living organisms are locally suboptimal: they make use of reactions, of enzymes and of processes in a non-optimal manner, meaning that these players are often redundant, relatively inefficient, heterogeneous, random or incoherent. And yet the integration of all of these imperfections, at least from the perspective of the 21st-century human being, gives rise to adaptable biological systems. In other words, life is built on the absence of local optima.

Suboptimality could thus be defined as the ability to evolve in the long term, utilising internal weaknesses, not as problems that must be circumvented, but rather as springboards enabling adaptability. Conversely, and in light of these internal weaknesses, suboptimality implies that the system is not functioning to its maximum capacity. It is therefore a true paradigm shift: while optimisation weakens the system, suboptimality uses fragility to build systemic robustness instead.

Suboptimality is in fact designed on a population-based scale: the survival of the group and its evolution are prioritised over individual comfort and performance. It is therefore a strategy of collective resistance, founded on individual deficiencies and vulnerabilities.

The Japanese art of kintsugi (related to wabi-sabi, the philosophy of imperfection) could be used to illustrate this paradoxical point: by highlighting the cracks in the ceramic with gilded joins, the object becomes even more beautiful on account of its localised flaws and its magnified history. In other words, suboptimality acknowledges that "to live" in fact means "to survive": living organisms are constantly subjected to constraints and instead of living to their maximum theoretically possible capacity, living beings live beneath their capacities to stay adaptable. They are suboptimal. Weaknesses are thus reappropriated into strengths, as pointed out by Michel Serres: "for [a system] to adapt to changes, it must be conceived and built with play, in the sense that cogs have play, or give. Any evolution can only be born of fragility".[1]

Do we have evidence that this theory provides an accurate description of living beings? Is it possible to demonstrate that biological systems are less than optimal?

4.10 THE EXAMPLE OF BODY TEMPERATURE

Let us consider the example of enzyme activity. These proteinaceous workers populate our cells and are responsible for the main cellular functions, such as the digestion of metabolites, the construction of cell structures and compartments, the movement of the chromosomes during cellular division, the contraction and migration of cells etc. When one places these enzymes in a test tube, in the presence of their substrate of preference, their activity can be measured (for example by measuring the degradation of the substrate if the enzyme performs a metabolic digestive function). Place a piece of bread in a glass and add your own saliva, and you will see the salivary amylase do its work, namely that of digesting the starch in the bread.

Since our body has a temperature of 37°C, one might assume that all enzymes would exhibit an optimum level of activity at this temperature. However, biochemical studies have shown that the majority of human enzymes have an optimal activity at temperatures close to 40°C. This implies that at 37°C, most enzymes are suboptimal – they do not function at full capacity. The temperature of 37°C is thus a compromise that allows our enzymes to function well, but not *very* well.

Knowing that body temperature fluctuates around 37°C, enzymes would be suboptimal to be more or less active depending on the context. For example, when we are asleep, our body temperature generally drops by 1°C (to 36°C), and this is accompanied by the slowing of the metabolism associated with sleep. Conversely, the intrusion of a pathogen can induce a fever. The resulting increase in temperature can increase the activity of certain enzymes and, in a manner of speaking, surprise the pathogen, in particular by boosting the immune response.

The example of fever allows us to measure the added value provided by the suboptimality of enzymes. Take, for example, the *nuclear factor kappa B (NF-κB)* signalling pathway, which plays an important role in inflammations in response to viral and bacterial infections. The proteins involved in this pathway regulate the expression of the response genes and stimulate the production of cells from the immune system. When the body temperature is at 37°C, this signalling pathway is only active to a minor degree, well below its activity maximum. This pathway is activated drastically when the temperature exceeds 37°C.[254] This response depends in particular on the production of a protein named A20, which causes the activation of the NF-κB complex when the temperature rises. For instance, when one artificially blocks the production of the A20 protein, the NF-κB complex is not activated when the body temperature increases. Here we see the benefit of a suboptimal temperature to this signalling pathway: not only does the rise in temperature increase the activity of the enzymes and proteins involved in resisting the pathogen but it also enables the amplification of the warning signal and of the subsequent immune response. If this signalling pathway were optimal at 37°C, the increase in temperature would not enable such a stimulation of the immune response. Importantly, this response cannot be maintained forever, since here performance gains become counterproductive in the medium term: knowing that most proteins become denatured above 41°C, fever can cause lesions if it is too intense or lasts too long.

The example of fever shows that living organisms can increase their performance, but that this situation can only be transitory. In normal conditions, the case of enzymes shows that suboptimality does indeed exist in biology, and that it even has benefits to our health by providing enough slack. Before considering other examples in greater detail, we must ask a question: Is suboptimality strictly confined to the biological world, or is it in fact applicable to other areas? Let us take the case of decision-making in the economic sector.

4.11 SUBOPTIMALITY AND VIABILITY IN ECONOMY

When we have to make a decision, we often consider several parameters beforehand. Such a theoretical approach is called the *multi-criteria approach*. It is widely used in the economic field. Here, I would like to briefly outline its limitations in order to arrive at an alternative that is closer to the bio-inspired suboptimality.

What is the principle behind the multi-criteria approach? This technique incorporates a large quantity of data and a large number of interactions and objectives. According to criteria defined by the user, it determines mathematically whether one action is better than another. For example, there are several different ways to travel to your place of work (on foot, by bicycle, using public transport, by car etc.) and there are several possible routes. Without necessarily formalising it, you probably adopted a multi-criteria approach to select the best way to make your first journey between home and the workplace. And you most likely gradually optimised it. For example, Londoners or New Yorkers might optimise this journey in choosing whether to wait for the subway train at the start or the end of the platform. Since one given solution will not satisfy every criterion (as a general rule), the aim of the multi-criteria approach is to identify an optimal solution. In order to make this choice, the user ponders the various criteria (social, economic, environmental) arbitrarily. Picking up once again on the example of the journey between home and work, your private car is perhaps the fastest way to get to work, however, this may come into conflict with your environmental and social ethics, which would instead favour public transport. This raises the question as to the, very much empirical and subjective, justification of such choices.[255] So how can we define an incontestable optimality criterion? Furthermore, the choice to use a private car involves a financial investment and is thus a choice made over a period of time. Within two years, the price of fuel could triple and, therefore, the choice made today could already be obsolete. How can we justify a criterion without knowing the future? How can we be sure that an optimum at a given moment will still be relevant in the future?

The concept of *viability* offers an operational response to the limitations of optimality driven by the multi-criteria approach. Exhaustive research on viability and its mathematical formalism can be found in the PhD thesis written by Claire Bernard.[256] Simply put, viability concerns the study of dynamic systems subjected to a selection of constraints. In contrast to the multi-criteria approach, it boasts the advantage that it identifies solutions without the need to prioritise the desired criteria.[256, 257] We can already see how this approach is adapted for living organisms, them being dynamic, horizontal and evolving systems within fluctuating environments. The semantic choice of the word *viability* is actually by no means trivial.

Viability identifies at least one evolution in which all constraints are indefinitely satisfied. This evolution is referred to as the *viability kernel*. What exactly does this entail? Rather than fixing the parameters once and for all, one instead fixes the parameter boundaries: evolutions are free, but remain channelled by a selection of constraints. This implies that the selection of variables is not made *a priori*; variables will instead adapt to the evolution of the system over time.[256]

It may be that no viable evolution exists. In this case, the viability kernel is empty. In order to find a mathematical solution, one must then modify the level of constraint, or the structure of the system. It is also possible that evolution will merely be impossible for a certain period of time, known as the *crisis time*. One of the objectives of the viability approach is thus to identify these crisis times, in order to minimise their duration.[258]

Viability has been applied in several fields. For example, in demography it has guided the choice as to whether or not to have children, taking into account numerous constraints such as the maintenance of a certain quality of life, or the ability to raise children.[259] Viability has also been used in biology to study the preservation of genetic diversity within a population.[260]

The case of atmospheric CO_2, and the capacity of forests to regulate it while offering vital economic opportunities, is interesting within the context of the Anthropocene question. Is it possible to identify a form of evolution that would be capable of satisfying the associated constraints (rate of deforestation/reforestation, level of CO_2 emissions, financial resources associated with forest products)? One study has shown that the viability kernel is empty,[261] and that the current system is therefore not viable! In the case of fishing, it has in fact been possible to identify viable evolutions.[262]

Our society is obsessed with optimum productivity, efficiency, competitive capacity and even security. Biology indicates that the unique optimum is an impasse, since it does not anticipate fluctuations within the system. In economy, the viability approach formalises a method by which to take into account the range of options when subject to constraint, and thus validates the strength of the non-optimum in responding to a problem in all its complexity.

So why don't we use the term viability rather than suboptimality? While the two concepts can be related, I consider it more pertinent to speak of suboptimality, since this term implies that optimisation poses a problem. The word viability might lead one to believe that a form of expanded optimality would suffice. Even worse, a misunderstood viability could lead us to reduce constraints rather than finding a path that accommodates and uses constraints. From this perspective, the use of the term suboptimality compels us to critically evaluate optimality from the inside out. It also unfolds the notion of robustness, the capacity of a system to maintain its stability (in the short term) and its viability (in the long term) despite fluctuations: to be robust, one needs to be suboptimal. Robustness *via* suboptimality is more engaging than the notions of reduction and frugality. Indeed, (uninviting) sufficiency and (arduous and often counterproductive) efficiency gains are unlikely to trigger a societal drive towards sustainability. In contrast, robustness provides a universal engaging path in an increasingly turbulent world, which should lead to sufficiency in the end, without noticing. In suboptimality, a sustainable collective behaviour does not emerge from the addition of optimal individual behaviours, but rather from the integration of non-optimal and contradictory individual behaviours.

In addition to viability in economics, a similar conclusion could emerge in an even more generic way by studying the behaviour of dynamic open systems with a mathematical lens, notably via cybernetics or control theory. Biology has the advantage of providing more concrete evidence, which is easy to grasp, and of opening up a reconnection with living beings.

4.12 FROM PARASITE TO SYMBIONT

Whether this applies to the climate crisis, biodiversity collapse, systemic pollutions or the associated social tensions, the level of fluctuation will increase significantly in the future. For the time being, our response has been to optimise our environment in the short term. We are pursuing the path of idealised optimality, with all of the risks that this implies. In contrast, biology indicates that living organisms have instead disregarded local efficiency in the short term, in favour of long-term global robustness. They have incorporated multiple constraints throughout the course of evolution, and this ontological memory allows them to integrate the long term into their everyday behaviour. In other words, living beings do not foresee, but they are prepared.

To keep a balanced view, one can find some hyper-efficiency situations that verge on optimality in nature. For instance, when the temperature increases and nutrients are abundant in water bodies following pollution with nitrate or phosphate, bacteria and cyanobacteria multiply and form huge populations known as *blooms*. These organisms proliferate by cloning themselves: they thus drastically reduce genetic diversity within their environment. While resources remain abundant, these blooms grow larger and form huge colonies, sometimes even reaching a surface area comparable to that of some countries. Satellite images of these coloured blots, spanning thousands of square kilometres in the oceans, are particularly striking. In most cases such blooms produce toxins, which limit competition for resources, thus reducing biodiversity even further. The system becomes simplified. This is indeed an optimal strategy for the exploitation of available resources. However, a population feeding on the same resource in limited quantities cannot continue to grow in the long term. As impressive as the scale of the blooms might be, equally striking is the speed at which the collapse of the population takes place: all members of the population are in competition for the same resource, which inevitably means that it is doomed in the long term.[263] As with fever, hyper performance in biology can only be transient. In the case of blooms, simplification, through the optimisation of the exploitation of available resources, is not robust. It enables the rapid exploitation of available resources in the short term, but does not allow for survival in the long term. Blooms behave akin to gigantic parasites.

Humans are exploiting the resources of their host, the Earth, with ever-greater efficiency and speed, and without necessary worrying about their survival. Humans increasingly resemble blooms. The comparison could perhaps be broadened to include the toxicity of humans towards other species, with humans alone capable of bringing about the sixth mass extinction of species. Thus, the bloom analogy offers a rather simple definition of the Anthropocene: the period during which humans behave like parasites towards the Earth.

This analogy is pertinent in other regards: it highlights the limits of survival strategies within an open or closed system. Within a spatially and temporally open system, as is the ocean ecosystem to the bloom, the short-term strategy is ultimately the winning strategy: the parasite can develop quickly and produce abundant offspring capable of waiting out the regeneration of the ecosystem in the subsequent cycle, in the form of spores, for example. Within a closed system, this strategy would clearly be doomed, since the ecosystem would not regenerate after depletion. The Anthropocene is precisely the moment when humans realize that their domination on the whole planet confines them to a narrow world. Humans inhabit a spatially closed system.

Is there then a way out in the time dimension? In theory, humans could wait a sufficient length of time until resources have been recycled and regenerated. However, even in this case, we would essentially remain within a closed system, since the duration of human life is quite simply incompatible with the time required for the regeneration of non-renewable resources. For humans, the Earth system is a closed system, both spatially (we are exploiting the entire territory) and temporally (our lifespan does not allow us to wait for several thousand years in the form of spores or seeds). The human strategy of optimisation is the strategy of the parasite, but within a closed system. If we wish to coexist with the planet, we must therefore learn to become symbionts, that is, to coexist with the planet, adopting a circular approach. Life having long-term experience of symbioses and cycles, it might help us along this path. Suboptimality is the beating heart of these robust strategies.

5 Robustness of Life

It may be difficult to envisage a suboptimal third way without a precise knowledge of its underlying mechanisms. In this chapter, you will find a selection of examples of suboptimality in living organisms. In a civilisation that is drawn to optimality, this scenario would have little chance of attracting votes. And yet: randomness, redundancy, waste, fluctuation, variability, heterogeneity, slowness, hesitation, incoherence, imprecision, error or incompleteness contribute to the robustness of life.

I have drawn on a limited set of examples. This chapter by no means purports to be exhaustive: another biologist might have used completely different examples, ultimately arriving at the same conclusion. In parallel, you also have a right to non-exhaustivity: we now come to a section that is more biology-heavy, and you may not wish to read every detail. Each sub chapter being independent, this chapter offers the opportunity to engage in a more relaxed approach to reading: as proposed by writer Daniel Pennac,[264] you can pick a sub chapter randomly before moving to the last part. Perhaps, a form of suboptimal reading?

5.1 RESILIENCE VERSUS ROBUSTNESS

Before taking a detailed look at some biological examples, let us first come to semantics. In particular, you may be surprised to see the word "robustness" and not "resilience". Here I would like to explain this choice.

I have already defined robustness as the ability of a system to be stable (in the short term) and viable (in the long term) despite fluctuations. In other words, a system is considered robust when it is capable of maintaining its responses within a given range in spite of environmental variability. This is, for example, the case with human body temperature as discussed earlier: the human body maintains a body temperature within the range between 36°C and 38°C, in spite of temperature fluctuations from day to night, or from winter to summer. The idea of stability associated with robustness does not mean rigidity or a passive response to external disturbances. Rather, robustness involves an internal flexibility to actively resist fluctuations from the environment.

Robustness is very close to the notion of adaptability. A system is adaptable when it is capable of adjusting its responses while maintaining its central functions. This means that, for a system to be adaptable, there must be enough room to be sure not to fail. The human body, for example, is adaptable, since it increases its body temperature in the event of infection, in the form of a fever, while still maintaining its capacity to breathe or eat. Both robustness and adaptability play against optimisation: they require extra leeway to accommodate different and unpredictable events. They push boundary conditions within which different viable strategies can develop.

Resilience derives from the Latin word *resilire*, to bounce back. It was originally defined in material sciences as the ability for an object to deform and get back to its

DOI: 10.1201/9781003510918-5

original shape. Resilience was then applied to psychology where it refers to the ability to fall and recover (after a trauma, for instance). Is this the same as robustness? Here I argue that resilience is almost the opposite of robustness for two main reasons.

First, whether in material sciences or in psychology, resilience describes a stereotypical trajectory: going down and bouncing back. A trajectory, by definition, can be optimised. For instance, when falling, one could try to be down for the shortest amount of time and to bounce back as fast as possible. When having a cold, taking an aspirin to reduce the fever to get back to work as soon as possible could be seen as a combination of resilience and efficiency, for instance. In other words, resilience is compatible with the cult of efficiency through the most optimal path. In contrast, to be robust, one needs to have extra leeway. This involves some degree of inefficiency, to allow a diversity of possible trajectories. To continue with the case of the cold, one would instead stay in bed and let the fever go on (if manageable), to make sure the immune system learns from the pathogen attack, even if this requires extra days of sick leave. Robustness rather involves de-optimisation.

Second, whereas robustness describes conditions in which one does not fall, resilience builds on falling down. Resilience can even become an incentive to fail, to be able to learn from it. Taking again the example of body temperature, robustness would come with frequent exposure to small doses of antigens (vaccination) to avoid major diseases, whereas one would need to fall really sick to know whether they are resilient or not. Psychological resilience can make sense in very specific medical or social cases; it becomes extremely problematic when it is extended, and even imposed, to the entire society. This turns into a competitive credo that is perfectly aligned with the neoliberal economic model. In other words, resilience, as an incentive, can be seen as a form of extreme fatalism that delegitimates any form of resistance. To be convinced of how toxic resilience has become, one can explore Japan's national resilience policy after the 2011 Fukushima disaster. As related by social scientist Thierry Ribault,[265] this takes the form of a fatalist order: "to not be afraid of nuclear radiation, you have to expose yourself". In short, resilience is becoming a scam: it carries the positive idea of recovery, through an invitation to fall.

In a society that is obsessed with performance and personal development, this definition of resilience is dominating today. There is however an exception: ecologists have created their own definition of resilience, as the capacity of a system to be robust, adaptable and transformable.[266] The confusion with robustness is remarkable, and problematic. Another issue is the addition of the transformable component. This deserves a small explanation.

A system is transformable when it can evolve into a different system, by modifying its structure and its functions. As environmental scientist Carl Folke and colleagues wrote in a seminal article, "transformational change at smaller scales enables resilience at larger scales".[266] To continue with the analogy of body temperature homeostasis, let us consider its origin to see how transformability operates. Birds and mammals are the only truly endothermic organisms, that is, organisms capable of regulating their body temperature independently of external temperature. During the transition from the Permian to the Triassic, almost 250 million years ago, a +10°C increase in the Earth's temperature, itself caused by massive CO_2 emissions of volcanic origin (Siberian and Emeishan traps), leads to the disappearance of

96% of marine species and 70% of continental species. This extinction would have been spread over tens of millions of years, the most notable consequence being an 80% drop in the oxygen content of the oceans.[267] Endothermy may have afforded pre-mammals and pre-birds an evolutionary advantage: this transformation would have enabled them to survive more effectively than other species.[268]

Including transformability within the definition of resilience is very confusing. Resilience in material sciences and in psychology is all about conservative features, and not about transformation. To avoid confusion, I would have rather written that "transformational change at smaller scales enables viability at larger scales", thus a question of long-term robustness. Finally, when people talk about resilience, without defining it, it is impossible to know what they talk about. These are perfect conditions for neoliberal misappropriation and increased responsabilisation of citizens: if you are not resilient, it is your fault because you were not strong enough to get back on your feet. Robustness on the contrary implies that other options are available to face and survive fluctuations, to make sure you do not fall. In the long term, by authorising different trajectories within certain boundaries, a robust system can evolve over time to remain viable.

Distinguishing resilience and robustness is also necessary to clarify the definition of other words that are often misused in the socio-ecological and economic fields. Antifragility, for instance, is close to psychological resilience, with the idea to bouncing back after a shock at an even better position than the initial one. Antifragility can apply to certain biological features. For instance, a broken bone often becomes stiffer after recovery. In other words, antifragility describes situations where "what doesn't kill you makes you stronger". When this works, as in bones, this can be positive. But, the critic of psychological resilience also applies to antifragility with the same rationale: it is an invitation to fall to become even more performant. And the success of the antifragiles only applies to the survivors. For the others, usually the vast majority, "what doesn't kill you makes you sicker".

Another example is the word "agility". It is the ability to slalom between the risks and the fluctuations. It is perfectly aligned with optimality, precisely because the whole point of agility is to find the best and shortest path, between the fluctuations. To be agile, one needs to foresee the risks. The incentive to be agile is thus a hidden invitation to fail sometimes, simply because it is impossible to foresee all risks. Agility is perfectly aligned with resilience, in its extended psychological form. Adaptability can be opposed to agility, just like robustness can be opposed to resilience, because it implies different routes and some degree of inefficiency to maintain these alternatives. In other words, the invitation to be resilient, antifragile or agile can be fatal; the only way this does not happen is if you are robust and adaptable in the first place.

Now that I have clarified why you won't see the word resilience in this book anymore, let us explore the underlying mechanisms of the robustness of life, beginning with randomness.

5.2 RANDOMNESS

The aforementioned example of the flagellum found in bacteria could cultivate an engineering-based vision of biology, that is, a determinist system in which everything is predictable, as in a car engine. However, this masks the numerous random events behind the selection of this type of structure during its evolution, and the many unforeseeable factors that could alter its function within a highly fluctuating biological context. This is the definition of stochasticity: the impossibility of correctly predicting an action, on account of the system's internal and external unknowns, as in a game of dice. We thus talk of a probabilist approach, as opposed to a determinist approach.

Even if living organisms are not the products of an industrial design, the deterministic view of molecular biology remains fairly widespread, even in textbooks. For instance, cell receptors and enzymes are often shown as locks into which ligands and substrates fit perfectly, like keys. An array of educational films presents the life of the cell using operational sequences, on a par with modern day assembly lines. These didactic animations are by no means useless: they highlight certain interactions and help us understand the associated processes. While progress is made with a view to better representing molecular fluctuations,[269] these movies do not take full account of the stochastic reality of a cellular environment. For example, the movements of molecules within a cell generally result in collisions with other molecules, and not in the precise guidance of trajectories. Adopting a probabilist approach, we must therefore imagine that the key is held by the hand of a drunkard who takes several attempts before finding the keyhole, if he indeed finds it at all!

The consideration of unknowns in biology has given rise to a new field of research.[270] It involves understanding the conditions for the construction of living, self-organised, systems, adopting a compromise between determinism and indeterminism, "as if a certain quantity of indeterminism is necessary in order to permit a system to adapt itself to a certain level of noise".[271] This field of study, inspired by game theory,[272] has remained speculative for a long period of time. The emergence of new technologies capable of detecting single molecules in cells and tissues has opened up a new, experimental dimension in recent decades.

When the microscope resolution is low, one can only observe the overall behaviour of a population of molecules and not the movement of each individual molecule. This might lead one to naively believe that each molecule behaves like the overall population of molecules, like soldiers marching in unison. However, analyses of the trajectories of each molecule in the cell more often than not show highly varied paths, somewhat akin to travelling from Paris to Marseille, but passing via London, Madrid and Bengaluru.

To illustrate this point, let us consider the case of cars navigating a roundabout. Viewed from above, at a low resolution, this movement is purely circular and at a constant speed, however, viewed close up, at a high resolution, each car has its own trajectory and its own rhythm. It is indeed the increase in resolution that enables the detection of the noise, or the vagaries of each car trajectory.

Brownian motion provides an illustration as to how individual trajectories can affect the behaviour of the population, on a microscopic scale. In 1828, botanist Robert Brown observed pollen grain suspensions presenting small random movements in water. He did not manage to explain this phenomenon, however, later Albert Einstein (1905) and in particular the founder of cybernetics, Norbert Wiener, proposed that the pollen grain is continually bombarded with water molecules. The small size of the pollen grain means that it is occasionally bombarded by two molecules on one side and just one molecule on the other, which is the reason for its movement. If the pollen grain were bigger, its larger contact surface would average out these impacts and it would remain stationary.[273]

Here we find a typical effect of the law of large numbers, which is also evident in some surveys: the behaviour of a small sample of the population does not necessarily correspond to the behaviour of the population as a whole. This is one of the greatest sources of stochasticity in biology. For example, if there are 100,000 copies of a certain molecule in a cell, removing a dozen of them will not fundamentally alter the behaviour of the cell. In contrast, if there are only 20 copies of that molecule, the loss of 10 of them, that is half of the initial pool, will most certainly have significant consequences.

Do we know the number of molecules in cells? It has been estimated that more than 80% of the genes in the bacterium *Escherichia coli* each produce less than one hundred proteins per cell. Thus, of 2,100 proteins detectable on electrophoresis gel, only 300 to 400 genes are responsible for the production of 90% by weight of the bacterium's proteins. The 1,700 to 1,800 remaining proteins are thus in low quantity, ranging from just a few up to around 100 copies.[274] The number of molecules per cell also is variable, that is to say, there is a large margin of error concerning the average value per cell, the functions and fate of the cell will depend on these relatively few molecules and their fluctuations.

To consider another analogy, it is as if the number of assistants and chiefs of staff were to fluctuate within a government. Needless to say, information would not always circulate smoothly. One might rather determine the probability of receiving a piece of information based on personnel availability. It is the same in biology: we often reduce the function of a cell to an organigram, that is, the interactions between molecules, with activators and inhibitors, for example. However, in order to really understand how it works we must also include the possibility that a molecule might be temporarily absent, or in contrast, over-represented, that is the stochastic behaviour of these molecules.

What possible benefit could such uncertainty offer the cell? To answer this question, and before moving on to consider an example on a microscopic scale, let us consider a more immediately accessible macroscopic example.

Randomness is a key factor in determining the robustness of populations of living organisms. In fact, it enables the continuous preservation of genetic diversity within the population. In turn, this heterogeneity allows for selection during evolution. Individuals adapted to their environment will survive. Without the random mixing of genes and individuals, the population would run the risk of becoming genetically homogeneous, and that would undermine its survival. For example, if environmental conditions change, those fittest prior to the change could become the least fit after the change. The diversity of the population ensures the continuing existence of a pool of individuals resistant to new environmental conditions.

There is no lack of examples to illustrate this well-established conclusion. A classic case is the resistance of human populations to major epidemics: resistant individuals will always be found. The confusion of randomness and luck actually emerges in language, the French word "chance" serving as an acceptable translation for both

English terms, for instance. You could object that cereal monoculture is an exception: Aren't the individuals within such an agrosystem extremely homogeneous? And yet, grain yields have never been so high throughout the entire history of humanity.[275] By monitoring the environment of the cereals to the maximum possible extent, by means of ploughing, irrigation, the use of fertilisers and pesticides, cereals grow within a world that is by and large disconnected from environmental unknowns (drought, soil quality, pathogens etc.). Throughout the course of the green revolution, we have chosen to optimise seeds and agricultural environments in order to achieve standardised harvests, which are easy to gather using machines. We have made the process less random and more predictable, in order to reap the benefits of determinist agriculture at a reduced human cost. Domestication is above all a domestication of fluctuations. Needless to say, this model is coming to an end, with the use of pesticides being met with increasing societal opposition, the use of chemical fertilisers polluting water tables and the destruction of soils resulting in the sterilisation of arable land. The example of monoculture is therefore not a counterexample, but rather an illustration of the fragile model of control and optimisation against that of robustness, denying the fundamentally random nature of life. The determinist approach, in its simplest formalisation, is not adaptable, whereas the probabilist approach is, by nature.

Can this also be applied on a cellular scale? Can molecular randomness offer benefits to the cell? As for populations of individuals, whether humans subjected to a viral attack or stalks of corn subjected to drought, the cells are also subject to the constraints of a fluctuating environment. For example, populations of bacteria will likewise have characteristics that are to some extent random. It is precisely this stochasticity that will enable some of them to resist antibiotics. From the perspective of the population of bacteria, randomness is an excellent thing! The gene expression in each bacterium is to a large extent stochastic: clones of bacteria, with exactly the same DNA, cultivated under the same conditions, in the same Petri dish, exhibit significant gene expression variations.[276] This molecular randomness underlies cellular stochasticity and the adaptation of cells and organisms to environmental fluctuations. Similar observations have been made on different scales and in different tissues. The stochasticity of gene expression is a fundamental and defining feature of biology.[231]

Are there other advantages to be found in molecular randomness? In developmental biology, local variations in the quantity of proteins generate asymmetries: for example, cell A has more M molecules than its neighbour B at a given moment, t. A short time later, this is reversed and A has fewer M molecules than its neighbour B. Over a long period of time, these minor variations do not have major consequences: on average, cells A and B have roughly the same number of M molecules. In contrast, if cells A and B were capable of detecting differences in the concentration of molecule M, and were able to amplify such differences, then a minor, initially random difference could become a major difference, and even the starting point for discrimination between two populations of cells, with different identities. Is this scenario purely theoretical?

Feedback such as this has indeed been discovered in living organisms. Cells acquire their identity thanks to regulatory proteins that are now well known. For instance, in animals, the *Notch* protein localised on the cell membrane splits into two pieces when activated, and the intracellular portion enters the cell nucleus to modify gene expression and thus alter the cell identity. One might therefore consider this a decidedly determinist path: an external signal determines the identity of the cell *via* the *Notch* factor. In reality, multiple instances of feedback render this mechanism considerably more adaptable. For example, in the case of the acquisition of neural identity in the fruit fly, it has been shown that cells that initially have a lower level of *Notch* have a greater probability of becoming neuron precursors: the fate of these cells initially depends on a random component. How can this be explained?

As in the case of the M molecule mentioned earlier, determining the fates of cells A and B, multiple instances of feedback between *Notch* and other regulators, such as the *Delta* and *Shaggy* regulators, amplify the initially small fluctuations, resulting in the ultimate segregation of cell identities.[277] Here it is apparent that evolution has selected a process for the definition of cell fate based on initial random differences. One of the added values of a probabilist system such as this is that, at the outset, all of the cells are potentially capable of becoming neurons. The final selection is made very gradually, by considering the balance to be achieved with the remainder of the cell population. A determinist process would have closed the door to this flexibility at a very early stage, with the risk of ending up having no neurons at all.

Finally, the consideration of randomness in biology is fundamental if we are to understand how living beings remain adaptable. One of the pioneers of cybernetics, Heinz von Foerster, even outlined a *principle of order from noise*. Hence, "it is favourable to have some noise in the system. If a system is going to freeze into a particular state, it is inadaptable and this final state may be altogether wrong. It will be incapable of adjusting itself to something that is a more appropriate situation".[278] This summarises the major contribution of randomness to the suboptimality and robustness of complex systems, living or otherwise.

Stochastic computer modelling of the living being is increasingly being used to explain certain biological behaviours, and it often does a better job than determinist models would.[279] For example, in genetics, stochastic models shed light on the processes of the attenuation and amplification of variations in gene expression,[280] on the sudden appearance of gene expression peaks,[281] on stochastic resonance (or how the addition of noise can make a weak signal detectable)[282] or on the phenomena of oscillation.[283] The probabilist nature of biology is also verified on other levels, from the population of molecules to the population of organisms. In all of these examples, stochasticity is not a handicap; it is a vehicle of long-term adaptability.

At this stage, one might have the impression that biological systems are so random that they are becoming too ineffective. For example, if the frequency at which buses pull up were truly random, you could wait for a week without a single bus leaving the depot, followed by a day with a bus arriving at your stop every 2 minutes. Complete chance is not optimal, both for users of public transport and in other contexts. Yet it can be observed that living organisms are, in contrast, suboptimal: they manage to get close to an optimum, which could be achieved with a determinist approach, while incorporating a high degree of stochasticity. Living systems are therefore adaptable, on account of their probabilist means of functioning, and also end up exhibiting a quasi-optimal global behaviour. So how do they manage to have their cake and eat it?

The solution can be found in the level of constraint applied to chance. What does this mean? The interactions between the elements within the system mean that nothing ever comes completely down to chance. Considering once again the example of the bus, by adding the constraint of its size and the speed at which it is travelling, it becomes physically impossible to have a bus at the stop every second. The size of the bus is a constraint on the frequency at which it passes by. The same goes for biological objects, where the cell cannot synthesise an infinite quantity of molecules: biological objects are subject to the laws of physics, and are therefore subject to constraints. We will consider other examples of such constraints on the following pages, and we will even see that the balance between randomness and these constraints fuels autonomy, a key feature of robustness.

However, there is one constraint that is specific to life: biological systems are reactive systems, that is, they actively modulate feedbacks. This involves natural selection-type reactions, but this time on a molecular scale. For example, the molecular assemblies with the strongest interactions could be better preserved. The book by Jean-Jacques Kupiec and Pierre Sonigo, *Ni Dieu, ni gène* (Neither God, nor gene) provides a number of enlightening and provocative examples.[284] Therefore, by combining the constant random exploration of what is possible (and therefore the ability of systems to adapt to any situation) with the addition of constraints to these possible scenarios, after the fact (interactions between elements, natural selection, redundancy etc.), we end up in a situation of high adaptation to the environment and high adaptability to related fluctuations. Suboptimality transforms randomness from a weakness into a condition to be adapted and an ability to become adaptable.

5.3 REDUNDANCY

Redundancy is the enemy of efficiency. For instance, we are often tempted to eliminate duplications within administration. When companies merge, departments disappear on account of becoming "redundant". Within a society founded on optimisation, there is no place for redundancy. It should, however, be noted that we still retain a degree of redundancy in certain cases. For example, hospitals all have a power generator that can take over in the event of a power cut. The automatic flight control system (AFCS) of airliners is usually composed of three independent modules, each with a different design: redundancy and heterogeneity are in place to avoid failure.[285] Redundancy is also present in social systems. For example, democracies have put into place protocols to guarantee the continuity of power in the event of an accident affecting the head of state. We thus optimise up to a certain point. In fact, our societies do not target maximal efficiency. They retain some degree of adaptability. The examples cited earlier do, however, appear rather specific. Does redundancy have a more fundamental, broader and more systemic function?

Biological systems are hyper-redundant. Think of trees and their many leaves or your own lungs and their many pulmonary alveoli. One of the reasons why randomness can be so widespread is the balance between randomness and redundancy. In other words, living beings use both randomness and redundancy to exhibit a wide spectrum of behaviours in order to adapt to external conditions in the long term. Such intrinsic autonomy feeds robustness.

What are the main determinants of redundancy on a molecular scale? The majority of multicellular organisms are diploids: there are two copies of every chromosome in each cell. This is a first security: even if a gene is absent or is not functioning within one chromosome, the other chromosome will grant access to a full repertoire of genes.

Genetic redundancy goes well beyond this. As a result of progress in the field of genomics, it has now been established that genes are often present in multiple copies within their chromosomes. This is coined as "multigenic families". More than this, genes belonging to groups or families of different genes can also perform the functions of neighbouring genes or compensate for the absence of a gene from another family. For instance, the case of innate immunity in mammals, with their selection of cytokines, chemokines and receptors with overlapping and redundant functions, ensures a stereotypical and reproducible response in spite of environmental or genetic variations.[286] It is as though a hospital were to have several hundred backup generators, powered by various different energy sources: fuel, wind, solar etc.

Redundancy is also structural, especially when considering the architecture of genetic networks. Redundancy develops on an upper level within the topology of the gene network. This redundancy is neither simple nor linear (as in the example of the power generator), but instead resembles a web in which each gene is linked to several others. This type of redundancy is a response to the increase in the number of interactions within a system, and is therefore consubstantial with complexity. It is similar to how, within a country, there are several possible routes by which to arrive at your destination, but ultimately, all roads lead to Rome. In biology, this cannot be reduced to the passive networking of redundancies. The form of the network can instead determine the function of the genes. In fact, the analysis of the gene network topology often provides more information regarding the exact function of the genes than the identity of the genes themselves. In other words, "tell me with whom you are interacting, and I will tell you who you are".

Conversely, analysis of the network topology also enables the identification of weak points caused by reduced redundancy. For example, such weak points correspond to important nodes within the network of genes. This is a bit like a major junction in a city, implying that the traffic on these routes is dependent on the status at a single point: a gene with multiple interactions, and therefore multiple functions, will be difficult to replace. The loss of these low redundancy elements can result in vulnerabilities to a large number of pathogens. This can also explain why the outcome of a benign infection is extremely variable, ranging from the pure and simple absence of symptoms to death.[287] In biology, the cases of low redundancy are the problems. To consider another, non-biological example, faults in our increasingly more electronic devices can be explained by the lack of redundancy in their design. Once again, optimisation weakens. It is, incidentally, here that the comparison between a brain and a computer reaches its limit[270, 288]: our brain suffers cell losses on a daily basis, but it still functions correctly thanks to redundancy.

While the network of genes or neurons enables the successful illustration of the fundamental role of redundancy in biology, this impact might seem confined to molecular or cellular aspects that are not readily understood in everyday life. Let us therefore consider a more macroscopic example. When we touch our noses, our arm makes a movement. There are several different possible trajectories. These different trajectories are enabled by redundancy: the presence of several joints make possible an infinite number of variants for this movement. In contrast, the optimised robot would only have three joints for this task and only one trajectory. While it is easy to understand that the increase in the level of redundancy enables the arm to serve as a versatile tool (i.e. a tool that allows us to do more than just touch our noses), are there other advantages to this redundancy?

In a robot, if each joint were to make an error, then the final result would be the sum of these errors, thus producing a trajectory that is potentially very different from that which was desired. The multiplication of joints and trajectories increases the possibility of corrections (feedback) and thus the possibility of averaging out. It channels the expected trajectory. This is the first adaptive advantage.

However, in the human being, one could imagine that evolution or education would eliminate these errors and variations, with a view to achieving a perfectly reproducible and predictable movement. This assumption is realistic: it is actually one of the goals of professional athletes who, quite literally, *perfect* their movements. Would redundancy therefore be useless in this case? The variation of movement is never completely eliminated. Even the most disciplined athlete continually performs varied movements, primarily due to the large number of joints and the associated high degree of freedom. In a typically suboptimal scenario, it is not the task that is optimised, but its variability that is confined to within a restricted space, through the interdependencies between movements and associated feedback.[289]

Therefore, there exists a range of possibilities, independent of the task to be accomplished, in which variability is constrained, or channelled. Redundancy such as this enables a form of adaptability, for example, if you injure one of the muscles or if the task to be accomplished imposes different requirements (e.g. if an object to be carried is heavier or lighter). Are there other added benefits to this form of redundancy?

Recent studies have shown that variability of limb movement positively correlates with motor learning capacities: the variability with which a person performs a movement is a good predictor of their ability to acquire new motor skills. In other words, the noise (in the probabilist sense) associated with movement is an essential component of learning.[290] In a study conducted by Puneet Singh and his colleagues, the subjects were required to perform a stereotyped movement from a point A to a point B with their arm. This took the form of a straight line (optimum). After several attempts, point B was displaced as the subject made their movement. This compelled the subject to correct their trajectory, which thus became a curve. By repeating the movement in the presence of the same interference, the subject learned to correct their trajectory and to improve the speed with which their hand reached point B. Inversely, when the interference disappeared, the trajectory was initially overcompensated (the trajectory was curved, whereas a straight trajectory would have been more efficient). This experiment thus enables the analysis of motor learning capacities and the establishment of a link between these capacities and the degree of redundancy. The authors were surprised to observe that the subjects (and the arms) that learned the quickest were those who had the most varied trajectories. On this basis, the authors propose that the difference between one's dominant hand (the right hand for right-handed subjects) and one's non-dominant hand (the left hand for right-handed subjects) is the presence of greater redundancy in the dominant hand.[291]

In other words, the additional degree of freedom provided by greater redundancy facilitates learning.

Although not necessarily in motor learning, it is well known that teaching is based on repetition and reformulation. This may open the way to a potential generalisation of this conclusion: motor and non-motor learning may require a form of redundancy. Conversely, it is apparent that the optimisation of a movement, as performed by an athlete, could limit future learning and adaptation. More broadly, a society that emphasises optimisation on every level, by avoiding the benefits of redundancy, not only deprives itself of robustness but could also condemn itself to a form of ignorance.

To conclude this part, let us return to Michel Serres's *The Natural Contract* and these few lines, indirectly praising the benefits of redundancy:

> We have made politics and economics into their own disciplines so as to define power: how are we to think of fragility? By the absence of a supplement [. . .] Strength has reserves at its disposal; it defends itself elsewhere, attacks along other lines [. . .] This explains why fuzzy sets, equipped with leftover spaces and refuges, can be so resistant. There's nothing weaker than a global system that becomes a single unit.

5.4 WASTE

Living organisms waste an enormous quantity of resources and energy. This comes as no surprise: randomness and redundancy are two causes of waste that can be inferred from the previous chapters. Given the added value provided by randomness and redundancy, this waste is not useless, but rather indirectly contributes to adaptability and robustness. One might argue that waste is the price to be paid in order to acquire a high degree of adaptability. In addition to the examples from the field of molecular genetics cited earlier, the energy metabolism of living beings abounds with examples of waste, in this case with the possibility of acquiring a direct measurement of the associated caloric waste.

Consider the most abundant, and certainly the most important, enzyme on Earth: the *ribulose-1,5-bisphosphate carboxylase/oxygenase*, or *rubisco*. This enzyme is responsible for fixing CO_2 during photosynthesis, thanks to its carboxylase function. Since carbonaceous biomass is the foundation of the Earth's biosphere, rubisco has a crucial role. A miracle of efficiency, perhaps? While most enzymes process a thousand molecules per second, rubisco binds only three CO_2 molecules per second. This very poor performance may in fact explain the abundance of this protein. A compensation allowed by a form of redundancy. In addition to this obvious lack of efficiency, you likely noticed that the full name of the enzyme ends with the "oxygenase" function. This means that rubisco can "go wrong" and process dioxygen (O_2) molecules instead of CO_2. Therefore, not only is rubisco inefficient, it is also not very specific. The alternative oxygenation pathway, called *photorespiration*, is extremely costly because of the resulting net loss of biomass, but also because many molecular responses need to deal with the consequences. In particular, these must detoxify the cell of ammonia. However, in doing so, photorespiration allows nitrogen fixation of nitrates. Once again, from an apparent weakness, in this case a huge waste of energy and metabolism, new skills are born. Clearly, evolution has made the best of this situation: for more than 3 billion years of photosynthesis, rubisco has not been optimised for better CO_2 fixation.

Note that biological waste is different to waste of human origin. Living organisms are indeed fully integrated with the Earth's biochemical cycles. This implies that local waste is always a resource for others. If we consider now the system as a whole, it would probably be more accurate to speak of recycling, or the circular management of resources. In the Anthropocene, humans have invented true waste: in contrast to all other spheres – the lithosphere, the atmosphere, the stratosphere or the biosphere, the technosphere is the only one in which recycling is, by comparison, largely absent. Is this entirely true? Can one find cases whereby the (non-human) biosphere has generated waste without recycling?

Geologist Peter K. Haff compares the absence of recycling in the technosphere to the widespread oxidation of Earth that occurred 2.4 billion to 3 billion years ago.[292] Let's play back the film. Four billion years ago, all living organisms are anaerobic, there is no or only little oxygen (*sensu stricto*, dioxygen: O_2) in the atmosphere. The ancestors of cyanobacteria "invent" photosynthesis, during the course of which water (H_2O) is split in two: on one side there are the protons (H^+) and the electrons that provide energy for metabolic pathways, and on the other side the oxygen is discarded in the environment. In an anaerobic world, oxygen is actually no more than toxic waste. One needs only observe the effect of oxygenated water on a textile or on the skin to understand how oxygen can cause significant damage. First of all, this oxygen pollution oxidated the rocks on the ocean beds and on the continents, which actually became red (green ferrous iron, Fe^{2+}, oxidises into red ferric iron, Fe^{3+}). It took 1 billion years to saturate these reservoirs. We are now 3 billion years back in time and oxygen is beginning to gently accumulate within the atmosphere.[293] The oxygenation of the atmosphere accelerated 2.33 billion years ago, which is referred to as the *Great Oxidation Event*.[294] About 2.1 billion years ago, the oxygen concentration in the atmosphere reached 4%. This was without doubt the first global atmospheric event caused by biological activity.

What was the response that emerged in order that the biosphere could survive within this now toxic environment? Two billion years ago, photosynthetic algae were producing ever-increasing quantities of oxygen. Yet in the meantime, oxygen waste had now become a fuel through the respiratory process: the mitochondria that populate the cells are little endosymbionts, of bacterial origin, which, rather than generating oxygen, actually burn it and even produce metabolic energy. Initial waste was the origin of one of the greatest biological innovations, which is still at work in our cells today, as well as in the cells of plants and fungi. The response of living organisms has not been to reduce oxygen production, but to use it as a new substrate and to integrate it into the Earth's cycles.

The generation and recycling of waste from the biosphere has fuelled evolution in the long term. In particular, the emergence of multicellular life could, in part, be a consequence of the energy manna provided through the use of oxygen waste in respiration. The atmosphere reached a peak oxygen concentration 300 million years ago, with an estimated value of 35% (compared to our current level of 21%). This coincided with the emergence of tree fern forests on the continents. In turn, this high concentration would also explain the large size of insects and amphibians in this era, and later the size of dinosaurs.

This evocation of the Great Oxidation and its response – the invention of respiration – echo the Anthropocene. Like Peter K. Haff, we can propose that the over-exploitation of resources by humans together with the absence of recycling in the technosphere lead to *Great Pollution*.[292] What is our response? Sustainable development, frugality and digital development would be more or less clumsy solutions to the prevalent non-recycling of the technosphere, on a par with the invention of respiration in response to the Great Oxidation Event. It will not have escaped your notice that it took 1 billion years to occur, and that innumerable species contributed to its coming about and subsequently to the management of the atmospheric oxygen. Humans in the Anthropocene have managed to change the atmosphere within two centuries, and they achieved this feat all on their own.

Looking at living beings, and their tendency to waste and dissipate so much, one conclusion comes: life manages abundance as a problem; thus, it wastes by default. Georges Bataille, in *La Part maudite* (The Accursed Share), provides a pertinent interpretation of waste in the living world: "It is not necessity but its contrary, 'luxury', that presents living matter and mankind with their fundamental problems. [. . .] The living organism receives [. . .] more energy [. . .] than is necessary to support life [. . .] a surplus must be dissipated through deficit operations [. . .] for living matter in general, energy is always in excess, the question is always posed in terms of extravagance, the choice is limited to how the wealth is to be squandered".[200] The very low yield of photosynthesis, whether in the capture of solar energy or in the absorption of CO_2 by rubisco, illustrates this perfectly.

For the sake of completeness, living beings have also developed strategies by which to manage scarcity. At this point one might mention slowed life (during hibernation or the formation of seeds or spores), the change of metabolism in mammals when fasting, and even sexual reproduction, which enables the exploration of a greater number of ecological niches by means of genetic diversification when resources become scarce.[295] A less well-known case is that of autumn leaves. When the cooler months arrive, resources become scarce and the tree begins preparing for winter. Rather than letting the leaves fall while green and loaded with metabolites, it will instead recover and store a significant portion in the trunk. It is actually the chlorophyll leaving the leaves that explains their yellow, orange and red autumnal colours. The remaining carotenoid pigments and anthocyanins left in the leaves only become visible when the chlorophyll is gone. In the end the leaves do indeed fall, however, this waste of organic material will not be lost, since it will provide nutrition for the soil humus at the foot of the tree.

While waste is shared among biological organisms in sustainable ecosystems and humans in the consumption society of the Anthropocene, the context is not the same. For Georges Bataille, the two world wars were not caused by scarcity of resources, but were rather a means by which to dissipate an excess of resources. Consistently, in the 20th century, richer countries were at war more frequently than poorer countries, whether by participating in the conflict directly or by supporting it from a distance.[35] Unlike other living organisms, humans have not found other means by which to dissipate their excess: "some wealth [. . .] is doomed to destruction [. . .], the possibility of pursuing growth is [. . .] subject to expenditure".[200] Living beings continually dissipate the excess and fuel the ecosystem and matter cycles. This surplus biomass makes it possible for ecosystems to function, and for species to evolve.

5.5 HETEROGENEITY

Redundancy guarantees a certain degree of homogeneity: by multiplying the genes, redundancy acts like a safety net. It ensures the function of the gene networks in the event of a one-off failure. In contrast, randomness would increase the degree of heterogeneity, on account of the unpredictable nature of the number of genes expressed at a given moment. The degree of heterogeneity that results from this dynamic balance is a central question in biology. As with randomness, one might be excused for thinking that heterogeneity is merely white noise, that is to say, an annoyance without consequence and without any particular function. In order to explore this point more closely, I am now going to present one of my fields of research that establishes a link between heterogeneity and suboptimality.

Working alongside my colleagues Arezki Boudaoud and Dorota Kwiatkowska, we analysed the heterogeneity of growth in plant tissues. Initially we assumed that neighbouring cells, growing at different speeds and each glued to one another through their walls, would do anything possible to find a compromise so as to resolve any mechanical conflict. It is similar to if the person beside you at the cinema were taking up the entire armrest, and even half of your seat. A negotiation would have to take place in order to return to a more balanced situation. To explore this assumption, we considered a mutant plant that is less able to react to this type of mechanical conflict. We imagined that in this mutant, conflicts would be amplified and growth heterogeneity would be greater than in a normal plant. This is similar to if you were too shy to say anything to the person beside you in the cinema, and that person would end up using half of your seat. We were surprised to observe the exact opposite: in the mutant, growth was more homogeneous. The mutant was actually more capable of finding a compromise than the normal plant. This suggests that mechanical conflicts between cells that are growing at different speeds are not resolved by means of a compromise, but that these conflicts are, on the contrary, generally actively amplified.[296] Using computational modelling, we further demonstrated that there does indeed exist an optimum for growth homogeneity, but plants do not reach that optimum, and thus behave in a suboptimal fashion. They even find ways to fuel growth heterogeneity.

Wherein might lie the evolutionary advantage of the active maintenance of a certain degree of growth heterogeneity in tissues? Given that the emergence of a new organ, such as a leaf, requires an initial growth differential, we propose that a certain level of growth heterogeneity confers a continual ability to form new organs. In other words, the tissues maintain heterogeneous growth, so that all cells are ready to initiate a new organ: instead of creating a difference in growth rate *de novo*, tissues simply increase a pre-existing growth differential.

Mechanical conflicts associated with growth heterogeneity are also present in animal organs. This has been shown in several studies focusing on topics ranging from the wings of a fly[297] to the valves in the heart of a fish.[298] In plants, it has been possible to comprehend the complex shape of the snapdragon petals by identifying a succession of mechanical conflicts between areas growing at different speeds or in different directions, and their resolution through buckling and folding events.[299]

Some molecules used by the cell in order to perceive the forces and deformations resulting from growth differentials have been identified. For instance, in plants, cells resist the forces by weaving stiff cellulose fibres into their walls. The molecular loom that achieves this feat is guided by filaments, the direction of which is determined by forces. In the case of mechanical conflict, the cell changes the direction of these filaments, and the cell is then mechanically reinforced in the new direction of the force.[300] It resists. This response can also explain how growth differences within plant organs, and associated mechanical conflicts, determine when organ growth stops. It also explains the variation in terms of how pointed certain organs are, such as in flower sepals, for example.[301] In animals, other molecular pathways are involved, but with comparable conclusions, such as those observed in the fly wings.[302]

Growth heterogeneity has other fascinating properties. If an organ or a tissue grows in a homogeneous manner, no mechanical conflict occurs. This implies that the cells have no mechanical clue as to the condition, size or growth of the tissue that they form a part of. Growth homogeneity makes cells blind, mechanically speaking. If growth is heterogeneous, cells, in contrast, have a great deal of information regarding the growing organ, provided they are able to perceive the forces resulting from the mechanical conflicts. By extrapolation, we can speak of *proprioception*, or the perception of the self. The cells read mechanical conflicts, which themselves reflect the status of the tissue, such as its size, its shape or its growth. In the aforementioned case, the signal that is read is a force, however, other mechanisms with molecular heterogeneity, and therefore a form of biochemical proprioception, exist.[303] Finally, growth heterogeneity provides an intrinsic awareness of shape and deformations, another cue for autonomy and therefore adaptability.

Thus, growth heterogeneity is instrumental for the development of living organisms: it underlies the emergence and the diversity of organ forms, and the associated conflicts trigger a proprioception pathway that channels their final shapes. In other words, heterogeneity generates a rich source of information for adaptable growth and development, whereas homogeneity is intrinsically poor in information and poorly adaptable, like a rigid program. Thus, heterogeneity in biology is far from white noise. It is a rich source of information and coordination.

5.6 FLUCTUATIONS

A fluctuation can be defined as a spontaneous deviation from the mean. In the most general case, fluctuations occur randomly and form successive variations in opposite directions over time. For instance, we have fluctuating moods, we are hungry and then we feel full. As for heterogeneity, the fluctuation is a key condition to the adaptability of organisms to their environment. Adopting a rather reductionist view of happiness, an animal that was constantly happy would not make any effort to feed itself or to reproduce. It would thus be condemned by competition from others. It is the ephemeral, and therefore fluctuating, nature of happiness that makes possible its survival.[54] While it is easy to understand how fluctuations can feed adaptability, it is much more complicated to grasp how an object can be both fluctuating and reproducible. Yet this is the case for most living organisms.

Consider your hands: they mirror one another in terms of shape, and yet the lines on your right and left hands are different and cannot be superimposed onto one another. If you had a microscope, neither could you find the counterpart to each cell of your right hand in your left hand. Likewise, the right and left arms are generally equal in length, with an average margin of error of 0.2%.[304] Tissues, organs and living organisms are more often than not variable on a cellular level, and yet they have shapes that are, on the whole, similar.

The presence of reproducible biological shapes is the foundation of taxonomy: different species are morphologically recognisable. Linné even defined them on this basis, providing the foundations for botanical and zoological classification. From a more functional perspective, having a given form within a species is an initial barrier by which to limit crossover between the different species. Other barriers exist that ultimately ensure that offspring have a balanced and comprehensive genetic make-up, with a complete set of maternal and paternal chromosomes and without any deficiency in certain genes and functions.

On top of this, having reproducible shapes is also important for interactions between individuals, such as the act of living in a colony. Try to imagine the architecture of an anthill if ants had highly variable sizes! Likewise, examples of co-evolution, such as the pollination of flowers by insects, require rather strict morphological criteria in order to produce an effect superior to that of pure chance. Here we find an example of external constraint channelling biological randomness, this time during evolution.

Within an industrial society, not only objects with reproducible forms are produced for a variety of reasons but the process for the fabrication of these objects also involves a strict and reproducible protocol, notably in assembly lines. Quite the reverse is true for biological objects: they are constructed by means of intrinsically fluctuating mechanisms; the interactions, feedback and external constraints make them reproducible *a posteriori*. What are the associated mechanisms?

The management of fluctuations can be passive. For instance, living organisms practice spatial and temporal averaging. In other words, if the cells have a certain form at a given moment, they will have yet another form a little later. After several iterations, averaging out over the period of time, the cells fluctuate around a stable form. This is known as temporal averaging. By increasing the total number of cells, the weight of cells with aberrant shapes within the population of cells reduces. This is known as spatial averaging. Once again, we find the law of large numbers coming into play here: if a subgroup of cells appears variable on a local level, by including a larger population of cells over a longer period of time, a reproducible average behaviour emerges.

This might give the impression that fluctuation is a constraint that biological systems avoid through a smoothing process: they produce more cells (spatial integration) and they increase the duration of growth periods (temporal integration). Are there more proactive mechanisms that would promote cellular fluctuations in order to achieve reproducible forms?

Let us consider the case of the flower sepals in the *Arabidopsis thaliana* plant: these organs generally resemble little leaves around the edges of the flowers. They have readily reproducible forms, since they protect the flower bud. These contain the reproductive organs, where the seeds will be formed. Needless to say, the sepals protect the plant's most precious organs. Yet the variability in the size and growth of the cells is very large in these protective organs. So, what do sepals do in order to achieve globally reproducible forms? In theory, temporal and spatial variability could cancel one another out. In fact, if cells exhibit growth that is variable in terms of space, yet constant over time, minor growth heterogeneities at the beginning of organ growth will then manifest themselves through major differences in form. If the cells have variable growth, not only in terms of space but also time, these variables cancel one another out and the final organ achieves a reproducible form.[305] In other words, the morphological similarity of organs cannot be explained by a reduction in local variability. On the contrary, two types of variability, spatial and temporal, may balance each other out *a posteriori*. This mechanism enables the sepal to continually correct its form, thereby responding to external factors. Here again, two apparent inefficiencies, temporal and spatial variability, form a balance that fuels autonomy. In biology, maintaining variability thus remains non-negotiable if we are to enable adaptability. Fluctuation is therefore not, strictly speaking, a constraint to be avoided. It is rather in accepting and utilising all fluctuations that living organisms construct reproducible forms.

If the case of organ reproducibility may seem too abstract, here is another example that links fluctuation and robustness around a simpler question: Why are plants green? Sunlight is white, that is, it contains all the colours of the rainbow. However, only a fraction of this light, red and blue, is absorbed by plants. What is not absorbed, green, is reflected, which is why plants appear green to us. Incidentally, this helps to explain why the yield of light capture by plants is so low. But this explanation is not entirely satisfactory: Why would plants waste so much energy? Why shouldn't they take advantage of the entire light spectrum? A recent study shows that the optimal absorption of photons would require a stable flow of energy at the molecular level. In theory, this would be possible if cellular factors and light were always at steady state. However, everything is moving all the time, at all scales in a plant cell and light is constantly fluctuating too. Thanks to computer modelling, it has been shown that the absorption of light in red and blue is a good compromise to manage such fluctuations.[306] In other words, during evolution, wasting green light energy was selected to make photosynthesis robust. Plant evolution did not select maximal performance (i.e. to absorb all available light), but rather robustness (i.e. to absorb less light to be able to respond to fluctuations). The green colour of plants thus embeds a deep philosophical lesson; our green world is showing us the light!

The link to fluctuations is essential to understand the robustness of life, and its suboptimality. The *Law of Requisite Variety* formulated by William Ross Ashby, an advocate of homeostasis, could be its theorem.[307] In a seminal article, Henri Atlan made a simple and clear analysis[271]:

> Suppose that a system is exposed to a certain number of different possible disruptions. It has at its disposal a certain number of responses. Each disruption-response sequence places the system in a certain state. Of all the possible states, only some are "acceptable" from the point of view of the purpose (or at least the apparent purpose) of the system,

which could be its simple survival or performing of a function. Regulation consists of choosing from the possible responses those that will place the system in an acceptable state. Ashby's Law establishes a relationship between the variety of disruptions, that of the responses and that of the acceptable states. The variety of available responses must be as large as that of the disruptions, and as small as that of the acceptable states. In other words, a large variety of available responses is essential in order to ensure the regulation of a system that aims to keep the system within a very limited number of states, even though it is subject to a wide range of attacks. To put it another way, within an environment marked by diverse unpredictable attacks, variety in the structure and functions of the system is an indispensable factor in determining autonomy.

5.7 SLOWNESS AND HESITATION

With fluctuation, the significance of time emerges. Slowness and hesitation are another two crucial factors in the suboptimality of life. Before turning our attention to certain cellular mechanisms that rely on slowness in order to function, one need only look at the mechanisms of evolution.

It is over a long period of time that species confront their environment and are selected (all while retaining a certain degree of heterogeneity, randomness and variability, as seen earlier). The long term enables adaptability, as it allows plenty . . . of time. As a counterpoint, to illustrate how adaptability is constrained when time is short, consider the impact of humans on the biosphere. More precisely, when *Homo sapiens* progressively emerged in Africa some 300,000 years ago, they lived and evolved alongside other species that co-evolved with them. When humans began to migrate and colonise the entire planet, they found themselves confronted with animals that had never before experienced the violence and the efficiency of the humans' social hunting strategies. Practically the entirety of the megafauna (i.e. animals of which the body mass exceeds 44 kg) of the Pleistocene disappeared: the large land animals of the Australian territory disappeared 45,000 years ago. The same ecocide took place on the American continent 12,000 years ago.[308] The fact that we can still see giraffes and rhinoceroses today is due to the fact that *Homo sapiens* first appeared in Africa, enabling the slow adaptation of the local fauna to this new predator.

Let us return to the biological mechanisms involving slowness and delays. We have already seen how long-term integration enables the averaging of variables. Slowness is therefore an ingredient that contributes to the robust nature of life. The term temporal redundancy could be applied here: if a factor is lacking, all that is required is that one wait for a sufficient period of time for another factor to appear and fulfil the lacking function. Do delays have other functions?

Delays are ubiquitous in biology and, in particular, they explain oscillating behaviour. Biology is rich with examples of such behaviour, such as the secretion of hormones (menstrual cycle), our internal body clock (circadian cycle) or the division of cells (cell cycle). For a factor to have oscillating behaviour, it must, generally speaking, be controlled by negative feedback, that is feedback that counteracts the initial action. However, this is not sufficient to generate an oscillation: a delay is also required.[309] You can experience this type of oscillating behaviour on a daily basis while taking a shower: the water is too hot, so you turn the cold tap (negative feedback), however, a certain period of time is required (delay) for the water to become warm rather than hot, and it then generally seems too cold, resulting in a second instance of negative feedback, this time generated by turning the hot tap, and so on and so forth.

Oscillations have essential roles in the development of organisms. Let us consider the case of repeated organs. Plans for the organisation of animals are filled with repetitive structures, from the rings of the earthworm to the segments of the insect body. In vertebrates, it is a repeated structure that defines the group. It is now well established that the vertebrae are placed one after the other, in a repeated sequence, during the course of development. This iterative process involves not only an internal clock but also feedback and delay generating the oscillations, and the succession of vertebrae.[310] Evolution has thus made use of delays in order to repeat structures and to grow the body.

Delay-induced oscillations are also crucial factors in the adaptability of organisms to their environment. Let us consider the example of plants. One need only film their growth and play it back sped up, and more often than not, you will see oscillating movements in the leaves, the stems or the roots. Oscillations in the roots could allow them to explore greater volumes of soil or to avoid getting stuck when encountering an obstacle.[311] Perhaps less intuitively, the oscillations of the stems and leaves could also allow them to perceive their own shape, in a typical proprioceptive process. The teams of Bruno Moulia and Stéphane Douady have explored this question. When a potted plant falls onto its side, the initially horizontal stem will gradually curve, pushing against gravity. Yet when one films this disruption, the stem can be seen oscillating strongly at the beginning, the oscillation then gradually reducing. It has been proposed that the cells of the stem perceive changes in the curvature of the organ and actively resist these changes.[312] This is a typical case of dampening. Plants are thus undeniably capable of perceiving their own shape. They do this thanks to growth delay, in this case during oscillations brought about by the response of the tissues to gravity.

Delays have an even more generic role: they are essential to bi-stability. A bi-stable system is a system that can have two stable states, like a switch that is either On or Off. The case of oscillations mentioned earlier is the most common; however, there exist other cases without regular oscillations: the system can spend half of its life in On mode, and then the other half in Off mode. Thus, the differentiation of cells can be viewed as the transition from a pluripotent state (On state) to the differentiated cell state (Off state), as when acquiring muscle cell identity, for example. In the most differentiated cells, a return to the previous pluripotent state is not possible. How can delays contribute to such biological decisions? Let us consider stem cells, which shed new light on the evolutionary advantage of delay.

A stem cell is a pluripotent cell that is capable of differentiating to several distinct cell types, and that can also regenerate. In particular, stem cells are responsible for the diversity of blood cells (red blood cells, lymphocytes, platelets, etc.) and, more generally, in tissue regeneration. They express certain genes, which give them this unique ability. However, does a "stem cell"-type genetic programme offer a universally applicable explanation? One of the factors that explain the ability of stem cells to differentiate to several states, while maintaining an ability to preserve the stem cell identity, is thought to be linked to the transcription and conversion delays.[313]

When a stem cell receives a signal, for example, a biochemical growth factor that guides its differentiation towards "muscle" identity, the stem cells begin to synthesise proteins that will transform it into a muscle cell. Each gene will first need to be transcribed into messenger ribonucleic acid (mRNA) and then the mRNA will need to be translated into protein, which can take time at each stage. In the meantime, another contradictory signal appears: it is now necessary to differentiate into bone. The cell recycles the synthesised proteins and commences a new protein synthesis that will transform it into a bone cell, and so on and so forth. The stem cell does not manage to differentiate: it hesitates. Thus, the stem cell identity is in part defined by delays, which enable hesitation. In other words, competence emerges from slowness. It goes without saying that stem cells are essential to the life of all multicellular beings. Hesitation by means of delay is therefore another crucial factor in the robustness of life.

5.8 INCOHERENCE

The use of contradictory signals in stem cells to generate hesitation, and thus pluripotency, offers an introduction to incoherence in biology. Similarly, the inconsistency of the rubisco, and its two contradictory activities for and against CO_2 fixation, has already been presented. More generally, with progress in computer modelling, we are now discovering how molecular networks are filled with incoherence.[314] For instance, certain factors can induce the expression of a gene *and* the expression of the repressor of the same gene. This is known as an "incoherent feedforward loop". Far from being exceptions, they are found everywhere in the molecular networks of life, including in the previously discussed stem cells.[315] Given the contradiction induced by such loops, one might assume that the final message would be garbled. On the contrary, this type of loop adds a great wealth of variation to the means by which gene expression is regulated.

Let us consider an example: the growth of roots allows plants to anchor themselves in the soil, but it also allows the root to explore the soil in search of areas that are rich in water and nutrients such as nitrate. The growth of roots is modulated by several hormones, one of which is auxin. The presence of nitrate in the soil triggers the expression of an auxin receptor, which inhibits the growth of the primary root and promotes the growth of lateral roots. The evolutionary advantage here is clearly apparent: the root stops growing wherever there is nitrate, and young lateral roots colonise a nutrient-rich environment. However, the complete cessation of the growth of the primary root could pose a problem with regard to the future of the plant. It has been shown that nitrate also induces the expression of a microRNA molecule, which results in the degradation of the auxin receptor. The cells would thus become insensitive to auxin. This is a typical case of incoherent loop, since nitrate induces two opposite responses (activation and inhibition of the auxin-signalling pathway). However, the activation of the auxin receptor precedes the expression of the microRNA. This delay allows the receptor to do its job, but only for a limited period of time. In this case, the apparent incoherence with regard to auxin signalling is resolved temporally: the coupling between incoherence and delay generates a transient response of the root to nitrate.[316]

Many other effects of incoherent loops have been demonstrated. For instance, the predictions of modellers, validated by means of *in vivo* experiments, show that this type of loop can accelerate the initial expression of genes.[317] Another study has shown that these incoherent loops enable the cells to detect the concentrations of biochemical factors, not in terms of absolute values, but in terms of their relative quantity, that is to say, they can detect the dynamic of their expression.[318] It becomes apparent that living organisms have transformed incoherence into something of a Swiss Army Knife.

In addition to the examples cited earlier, incoherence could have a key role in the autonomy of living organisms in the face of environmental fluctuations. Let us consider a case of incoherence in immunity. One and the same molecule, cytokine IL-2, can have opposing effects: it can induce the death or instead the proliferation of certain immune system cells (the T lymphocytes). IL-2 is therefore an incoherent signal, since it has the power of life and death. Is there an associated evolutionary benefit to such incoherence? One study proposes that the dynamics of death and proliferation induced by IL-2 are different, the former being linear and the latter cooperative. Thus, a single molecule, such as IL-2, could have two different functions depending on the cellular context. This property is essential in order to maintain cellular homeostasis, that is, a more or less constant number of lymphocytes. Cells have a tendency to die when they have a low or a high concentration of IL-2, but tend to proliferate with intermediate concentrations of IL-2.[319] So, what possible function might this double agent serve? Why not simply have one molecule that activates cellular death and another molecule that activates cellular proliferation? Modelling indicates that the two-molecule scenario ultimately only provides the options On or Off: the cells will all die, or will all proliferate. When the two activities are borne of the same molecule, even though the functions are contradictory, this enables the system to exist in two states simultaneously: death and proliferation. The system can thus exist within a wide range of parameters, and therefore within a wide range of environments. In a manner of speaking, this amounts to a cellular version of Schrödinger's cat paradox. More pragmatically, this property is particularly crucial in avoiding the scenario whereby the quantity of lymphocytes is entirely dependent on environmental variations.[319] In this case, incoherence serves as a shield protecting against external fluctuation. Protection emerges from internal autonomy, which emerges from incoherence.

These counterintuitive findings may appear very much specific to molecular genetics. However, the different patterns that are observed in the networks of genes, such as these incoherent loops, are also observed to varying degrees within the neuron networks, in the metabolic networks or in ecosystems.[320] The associated properties are therefore very likely also applicable to these various different contexts.

The case of incoherent loops also represents an opportunity to take a step back and reconsider the previously mentioned mechanisms behind suboptimality. We have seen that living organisms are built on randomness – a weakness in the eyes of the 21st-century human obsessed with control – and that this weakness is counteracted by redundancy, another apparent weakness as far as humans are concerned. Similarly, the shape of organs can emerge as a result of balance between the spatial and temporal variability of cell growth. Consequently, it seems that incoherence is a transversal factor for suboptimality: not only do living organisms build their adaptability on their imperfections but they also develop new properties by putting them in contradiction. Living organisms use conflicts to create a dynamic internal balance, and therefore a form of autonomy, during their development. In other words, autonomy emerges from internal contradictions, and robustness emerges from autonomy.

We could go one step further, this time taking a detour through the field of mechanics. It is now well established that most biological structures are in a state of mechanical equilibrium between tension and compression. For instance, cells have a membrane that is under tension and a cellular content that is under compression. This is caused by the presence of osmotic pressure (because it is rich in salts and osmolytes, the cell tends to attract water by osmosis and thus to swell) and, in the case of animal cells, to an additional capacity for molecular contractility under the cell membrane. Mechanically, cells are a bit like balloons whose envelope (latex) is stretched and whose content (a gas) is compressed. An equivalent situation exists for the aerial organs of plants where the epidermis is under tension and internal tissues are compressed. For instance, when you cut the surface of a tomato with a knife and it immediately opens, you reveal the presence of tension in its epidermis. We speak of pre-stressed structures: despite their apparent stable shape, these biological structures experience a permanent internal mechanical conflict.

These properties are well known to engineers. Typically, suspension bridges are also pre-stressed: towers are under compression and cables are under tension. Architects are fond of this type of design, especially since these structures have a certain mechanical autonomy. Thanks to their internal mechanical balance, pre-stressed structures better resist to fluctuations in the environment, such as wind. With incoherence, we come to another key point of biological suboptimality: evolution uses conflicts – randomness versus redundancy, spatial variability versus temporal variability, tension versus compression – to create autonomy, which in turn feeds the robustness of life.

5.9 ERRORS AND INACCURACIES

As we have seen, living beings contain many inefficiencies and build on them in order to fuel robustness. This naturally leads us to consider the role of errors. At first glance, mutations could be viewed as "errors" in the gene pool. External factors such as UV radiation or mutagenic molecules (such as benzopyrene in cigarette smoke) can modify the DNA. Built around an alphabet of four letters (known as nucleotides: A, G, C and T), DNA contains genes that could be compared to words constructed using this alphabet. Genes are, nevertheless, much longer than words. The average size of a human gene is 27,000 letters.[321] For example, here is a very small section of the human insulin gene: . . . CAGCCGCAGCCTTTGTGAACCAACACCTGTG . . .

A mutation in the DNA is like a drop of ink falling onto a page of text. If the drop falls between words, it is more often than not without consequence for the comprehension of the text. Now suppose that the drop fell in the middle of the word "contravention" and when reading the text you changed this word to "conversation". For the word (i.e. the DNA sequence), it is not of fundamentally serious concern – it is still a chain of characters. But for the meaning (i.e. the function of the protein encoded by this DNA sequence), it is problematic, since it can result in misinterpretation. It is also possible that your efforts to identify the word that is partially masked by the drop of ink might lead you to suggest the word "contvrtion", in which case the word itself and, at the same time, the entire phrase would cease to bear meaning. This can be equally problematic, if not more so. Finally, the new word might be a neologism, for example, "contaversion" – meaning a variant of a tale? In the same way, mutations within DNA sequences can be painless, lethal or debilitating, but they can also enable an increase in the range of competence of a species, and thereby enhancing its ability to transform during the course of evolution.

Mutations are not caused solely by external factors: our cells also make errors quite independently. For instance, the replication of DNA, which is particularly critical to prepare the transmission of the gene pool prior to cell division, is accompanied by several blunders. On average, the enzyme that duplicates DNA, the DNA polymerase, makes an error every 100,000 letters. In fact, the links between nucleotides in the DNA double helix are weak links, which, as their name suggests, allows a degree of wobble room. For example, DNA polymerase might, by mistake, position a T in the place of a C. Given that human DNA consists of 6 billion nucleotides, each cell makes an average of 120,000 errors prior to each cell division.[322] Within 10 hours, a proliferating bacterium will have created a colony of 1 million individuals, of which 10,000 individuals will have accumulated at least one mutation. Resistance to antibiotics is thus easily explained by such an error-prone predisposition.[322]

Luckily, one might say, there exist mechanisms by which to repair these errors. Only 1 in 1,000 mutations will remain as a permanent mutation, on average. The defective products of these mutations are also counter-selected. For example, if a mutation in the DNA leads to the production of an unstable protein, the cell will never see the product of this mutation. Likewise, if the protein renders the cell non-functional, this cell will generally be digested by the immune system. Obviously, these mechanisms do not always function correctly, and a small defect in the DNA may be the start of a severe illness. Because the majority of mutations are cryptic, this balance ultimately grants them all of the evolutionary benefits of the mutations (a reservoir of diversity for future evolutions), while minimising the disadvantages of the defects caused by the mutations (but without completely cancelling them).[323]

Are there situations whereby the living being actively promotes the errors, not for the evolution of their species, but to perform a physiological function? Some DNA repair proteins actually are double agents. This is the case with the Mfd protein in bacteria. Under normal circumstances, Mfd repairs DNA. But when the bacterium is under attack by antibiotics, Mfd instead promotes the formation of mutations in the DNA. The bacterium thus promotes its antibiotic resistance capabilities by increasing the number of errors.[324] In passing, the ambivalence of Mfd joins the long list of proteins with apparently incoherent functions in biology. Are there other mechanisms in which errors are stimulated for biological function, even in the absence of external perturbation?

In sexual reproduction, it is often said that children have certain characteristics of the two parents. It is less often said that they are the fruit of hundreds of mistakes. In fact, the formation of gametes takes place by means of a special form of cellular division, particularly subject to errors, known as meiosis. Immediately prior to this division, the chromosomes are actively subjected to major perturbations. In particular, one enzyme (SPO11) causes several hundred breaks in the DNA. Such breaks can occur in normal cells that are subjected to external factors, such as radioactive radiation. What is remarkable about meiosis is that, on the one hand, they exhibit a large number of breaks in the DNA, and, on the other hand, these breaks are induced and stimulated by intrinsic cell factors. Evolution has selected a mechanism that transiently tears apart the genetic material.

The current consensus to explain this phenomenon from an evolutionary perspective is that such DNA breakages stimulate a molecular pathway resembling the error repair machinery, and its ambiguities. In particular, given the number of breaks, the correction of errors is not complete: the repair process is fallible and even prone to more errors during the repair process, which involves DNA recombination. This explains why, following this genetic catastrophe, each chromosome is variegated, carrying the characteristics of both the father and the mother. Coming from this hybrid gene pool, it follows that the children will present a combination of the characteristics of the mother and the father. Evolution has thus selected a process based on the stimulation of errors and their partial rectification in order to generate genetic diversity within the population. In my attempt to explain the concept, I have considered the example of human beings, however, the same logic, with proteins that are incidentally very much homologous, can be applied not only to other animals but also to plants and fungi. The promotion of errors is a cardinal and universal value in life.

Poorly corrected meiosis errors may still send a chill up our spine. The level of control appears very low. And this is what is really going on. Whereas the arrival of a child can be decided, factors such as gender or eye colour are determined by a combination of the innate and the acquired traits. This is probably the most important choice that a couple has to make, and yet this choice is actually not a choice at all, since the level of control is so low: a child is the fruit of randomness and errors. It is also unique and different to its two parents. This carries the risk of producing a child that is less well adapted to its environment than its parents, should it suffer from a severe handicap, for example. This possibility is far from being negligible: congenital disorders occur in 3% of newborns in the United States, for instance.[325]

Sexual reproduction, complete with its randomness and errors, is a crucible for suboptimality. On the one hand, it promotes the robustness of the population: in promoting the random mixing of the genes, the population maintains a genetic heterogeneity; this enables the adaptability of a group of individuals to changing environmental conditions. On the other hand, sexual reproduction does not favour individual comfort: to have a child that is a clone of one of its parents would ensure the ability of each individual to adapt to the current environment, which sexual reproduction can never guarantee.

Here again, the tropism towards optimality in human societies emerges. We have already developed an array of tools with a view to practicing a form of covert eugenics. These tools are of varying levels of technical sophistication, and also varying in the degree to which they are deemed ethically acceptable. Nobody could cast stones at parents who only want but the best for their offspring, however, one must, all the same, understand the long-term implications of this important biological shift. In optimising our genes within the human population, through the reduction of the randomness and errors associated with reproduction at a given time, we will gradually loose our future capacity for robustness. For example, if given the opportunity to choose, the majority of parents would not want to keep an embryo affected by sickle cell anaemia. Yet this illness also has benefits, in that it brings with it a form of resistance to malaria. If we select the optimal human beings according to 21st-century criteria, without retaining these silent abilities, future generations may suffer.

Our randomness, brought about by error-mutations on a population-wide level, constitutes a shield against future changing conditions. Conversely, our choice to optimise the human being, under the illusion of error-free control, condemns us to tread the path of the total control of our environment. As Bruno David, head of the National Museum of Natural History in Paris, phrased it in an interview for the futurist magazine *Usbek & Rica*, "people no longer accept being vulnerable and thus seek to assign a responsible person when an event occurs that is seen as a failure".[326] We choose determinist optimality, and therefore the inability to adapt. On the contrary, living beings show us that robustness emerges when control is suspended. Almost a lesson of psychoanalysis.

5.10 INCOMPLETENESS

And to finish? Let us explore the virtues of incompleteness. Living organisms are continually replenishing themselves. From this perspective, they are perpetually unfinished. Possible synonyms for this adjective include imperfect, partial or even insufficient and dysfunctional. Once again, life embraces the values that human beings are inclined to judge as negative. Wherein might lie the benefit of incompleteness to the robustness of life?

Trivially speaking, the incompleteness of living organisms is one of their main strengths. It explains our ability to recover from superficial injury (the body being capable of regenerating tissue, notably thanks to stem cells), or the ability of a given plant to adapt its morphology to fluctuating or different environments. Conversely, completeness – a form of perfection – is death: to be complete is to be finished. We thus come back to the notion that life is in motion, and therefore perpetually imperfect.

How do tissues and organs manage to function well without being completed? As we have seen in the previous pages, living beings host random, heterogeneous, variable, incoherent processes. In such a context, completion is not on the agenda. But we have also seen that all these parameters antagonise each other, and ultimately create a dynamic equilibrium. This allows for proper functioning and strong autonomy in the face of external fluctuations. Incompleteness is therefore not a failure, but rather the marker of a continuous dynamics, essential for the robustness of the living. Finally, incompleteness emerges from the suboptimality of life. Too great a number of interactions prevent the achievement of any form of perfection. To some extent, there will always be a variable that will drive a cell towards an alternate destiny, or to behave differently to its neighbours, and therefore the order will never be complete. This vision is fairly coherent with the stochastic view of biology: minor disruptions are always occurring, which means that the system is never functioning optimally.

Research at the frontier between biology and physics goes further still. The incompletion of living organisms is consubstantial to their perpetual capacity to resist, and in particular to resist forces. In particular, according to Wolff's law, the internal architecture of bones is made of trabeculae oriented to resist the forces caused by body weight and locomotion.[327] In other words, the bones are built along the lines of force and in return, the internal pillars thus built play a role of mechanical support. A bit like in a building where the walls are vertical, primarily to resist gravity. This law emerges from a cellular response: cells perceive the mechanical constraints of their environment and completely modify their internal architecture to resist them.[328]

And in plants? As we have seen previously, plant cells weave cellulose fibres into their walls in order to mechanically resist their internal osmotic pressure.[300, 329] Bearing in mind that cellulose fibres are as rigid as steel, this resistance is a mechanically tangible fact. In fact, when one prevents plant cells from correctly synthesising cellulose, the cells swell and explode.[330] It is now well established that cellulose fibres are deposited in an orientation that allows cells and tissues to withstand mechanical stress. They are aligned with the maximum stresses.[300] The external skeleton of plant cells thus follows the same law as the trabeculae of our own bones. In other words, incompleteness creates mechanical stresses and conflicts. In turn, these constraints are perceived and this allows growth to be channelled. In biology, living organisms remain unfinished all their lives in order to have a dynamic perception of their own growth and to guide it.

However, incompleteness is not limited to growing tissue. Let us consider plants and their aerial organs. At first sight, we might think that incompleteness only relates to young growing organs: the plant would pile up structures one on top of the other, and only the end of the stems would be unfinished. The growth in width of the tree trunk shows that this is not the case. But another property interests me here. Considering the young plants, you can see for yourself that most plant organs are made up of linear forms (one-dimensional: stems, petioles, peduncles etc.) and flat forms (two-dimensional: leaves, petals, sepals etc.). Why would such simple structures have been selected during evolution? These geometries have very large degrees of freedom. For example, one can bend a stem or a leaf along several axes. You only have to look at a young plant in the wind to see this. On the contrary, if you take an object with a more complex three-dimensional structure, you will lose degrees of freedom. This is, for example, the case of corrugated iron, whose mechanical strength is largely explained by its complex geometry. Similarly, the snapdragon flower has a very complex shape, with very few degrees of freedom for its deformation. We therefore arrive at the idea that the simple geometry of most plant organs allows large degrees of freedom, even in older, non-growing, tissues. In a more physical sense, the symmetrical shape of plant organs brings them closer to critical points.

The impression of incompleteness only at the ends of the stems is, therefore, an illusion: all the architecture of the plant, past and present, participates in the adaptability of the organism to its environment. Evolution has selected organs with simple forms that maintain an ability to respond mechanically to fluctuations in the environment (the wind, for instance) and to fluctuations in the plant itself (growth fluctuations in the short term, modifications in architecture in the long term). In other words, the plant builds shapes that can fluctuate. In turn, the fluctuations provide the cells with a lot of information to modulate the plant's architecture (proprioception). This perpetual loop maintains the ability of plants to adapt to their environment and to their own form: the response of plants to fluctuations leads to the formation of organs capable of fluctuating.[331]

This biological tribute to incompleteness opens up many societal analogies. For instance, new urban planners are beginning to question the completion-oriented nature of urban planning, historically in tune with finished, turnkey designs. On the contrary, they emphasise the values of unfinished forms. For instance, architect Patrick Bouchain, who received the Urban Planning Grand Prize in 2019, works alongside local residents to create collaborative and democratic projects, "which claim their incompletion to make room for the unexpected".[332] It is notably in this space that we could reconnect with life and its essence.

6 A Counter-Model?

As Alexis de Tocqueville wrote in *Democracy in America*, "a new political science is needed for a world altogether new".[333] To what extent can the robustness of life also inspire us to coexist with Earth in the Anthropocene? The mechanisms of biological suboptimality are not, strictly speaking, replicable or reproducible. However, one might find therein a few ideas or, at the very least, the first building blocks of a critique of the tropism towards optimisation in our societies. Let us begin by turning our eyes to initiatives that are currently underway, and that are already ushering in the world of tomorrow, and perhaps its underlying robustness. In order to overcome the failings of the Anthropocene, it is no longer a matter of improving performance, but rather of supporting adaptability thanks to suboptimality. This is certainly one of the most inviting prospects for humanity. In this final chapter, I will explore some of the weak signals of this new robust world under construction through numerous examples in social systems. This counter-model promotes values that are deemed negative within the optimalist dogma, but claims to ensure adaptability in the long term. However, the holistic nature of this view runs the risk of a superficial entry into distinct fields of research, which have their own history and conceptual framework. The idea of suboptimality in social systems thus remains a very open avenue. With this limitation in mind, I hope instead that such an imperfect and incomplete counter-model will have the virtue of stimulating a multi-voiced debate on the share of suboptimality that our societies might discard or accept in the future.

6.1 TOWARDS AN AMBIVALENT, BIO-INSPIRED WORLD

Before exploring the many suboptimal initiatives in our societies, let us get rid of the simplistic idea of bio-inspiration as a natural and universal answer to our problems. In order to reconnect with the physical world that surrounds us, our civilisation is adopting the trajectory of circular bio-economy, that is, an economy mainly based on green carbon of photosynthetic origin and integrated in the natural cycles of the Earth. By leaning on biological concepts, such as recycling or regeneration, that economy gives more appropriate consideration to the finitude of resources. This is the *bio-inspired* world of tomorrow. This is undoubtedly a major step forward in our reconnection to life. Yet have we abandoned our tropism towards optimisation?

The petrochemicals sector is turning increasingly towards so-called green chemistry.[188] This form of chemistry utilises materials of biological origin rather than oil, and also uses living organisms to transform them with a view to generating molecules with varying levels of added value (fuels, cosmetics, biomaterials, drugs etc.). Thus, either the molecules are transformed by means of enzymatic reactions performed *ex vivo*, or the organism is left to perform the entire synthesis *in vivo*. Most commonly, bacteria, yeast or algae are placed in reactors with their substrate, and the manufacturer purifies the obtained products. It should be noted that these biosourced

DOI: 10.1201/9781003510918-6

molecules are not necessarily the final products. They may also be intermediary products that refineries then modify further. The agrobiobase[334] catalogues a total of 300 biosourced products. The United States is home to almost 300 biorefineries and 20% of the world's chemical products are to be biosourced by 2030. For example, the company Solazyme utilises oil-producing algae to generate biodiesel that can be used by cars and planes (with the support of the American army), and is now venturing into the production of food supplements for vegetarians. The start-up LanzaTech uses bacteria that are capable of transforming certain industrial gases (carbon monoxide, CO_2 etc.) into ethanol, as well as other molecules, such as the bio-isoprene used in the manufacture of tyres, thanks to a partnership with Michelin.

The use of reactors limits the direct impact of the synthesis of these components on the environment and natural habitats. It should nonetheless be noted that the biosourced products are not necessarily biodegradable or compostable (i.e. they fail to degrade by 90% within 6 months, as per the EN13432 European standard). Incidentally, as is the case for palm oil, the strategy for biosourced products is not neutral for natural habitats, but rather impacts them indirectly. In particular, bacteria, yeasts and other algae require nutrients in order to grow, and very often these nutrients are molecules sourced from the agricultural biomass (sugar beet, maize, rapeseed etc.). Indeed, a quick calculation highlights the limitations of this strategy: the petrochemical industry currently consumes the equivalent of 400 million tonnes of carbon per year. Yet it is estimated that global agricultural production could not produce any more than 300 million tonnes of carbon for green chemistry, once losses in the manufacturing chain have been taken into account.[123] This echoes the frequent criticism of biofuels, with its indirect effect on conflicts regarding the use of arable lands. Green chemistry is promising but it is still a niche industry.

In addition to the use of living organisms in a reactor, field-grown plants and forest trees are also called upon to play a more significant role in the future bio-economy. For example, certain properties of plants could be used to cope with the prospect of scarce mineral resources. In particular, plants are capable of concentrating the minerals from the soil, and it is then possible to extract them in plant ash. The term agromining is thus used.[335] The start-up, Econick, is already running an economically viable project for nickel.[336]

Natural biological molecules also offer prospects in the area of substitution: the multiple properties of secondary metabolites, which have been used for a long time in pharmacy, could be required to replace certain rare minerals. It has also been shown that onion skins can mimic the properties of mineral metamaterials capable of filtering certain acoustic wavelengths.[337] The simplest molecules of the primary metabolism could themselves also contribute to this green revolution. In particular, the mechanical properties of cellulose fibres could be used for construction, which could help address the scarcity of construction sand.[338] DNA and its four-letter code offer huge possibilities for the storage of computer data.[339, 340] These examples also illustrate how agriculture and forestry will inevitably play a growing and versatile role in the time ahead: our future resources, within a circular bio-economy, shall increasingly depend on molecules produced either directly or indirectly by photosynthesis. Again, this would imply that attention should be paid to the equilibrium of cultivated areas, with a view to these new services permitting the production of food, while preserving natural spaces and the associated ecosystemic services. However, this scenario would ultimately entail the further optimisation of the landscape.

There may be an alternative, non-exclusive approach, namely that of sobriety. We could decide to produce and consume less. Manufactured products could be less complex, an essential criterion for enabling their repair and the recycling of their components. Finally, the emergence of the circular bio-economy leaves us faced with two options. In the first scenario, given the resources required in order for it to be viable, the bio-inspired world under construction would also incorporate an extended form of sobriety. This would involve curbing the human demand for ecosystem capacities. In a second scenario, sobriety would be considered incompatible with the current infrastructure and economic model. The aim would be to allow our green resources to produce yields adequate to feed the current economic machine, if possible, without threatening ecosystems. The latter scenario is very close to business as usual. In the end, we would pledge for a further optimisation of living beings to meet our needs.

6.2 DOMESTICATION – HOW FAR SHOULD WE GO?

Although the circular bio-economy places us on the path towards a reconnection with the environment, at the risk of the further rationalisation of cultivated lands, spectacular advances in the field of genetics offer the possibility of the rapid optimisation of biological organisms themselves. This in fact amounts to the major acceleration of domestication.

The double revolution of transgenesis and genome editing enables the very targeted modification of genes. Reference could be made to second-generation genetically modified organisms (GMOs), in which the manipulation is so surgical that it leaves practically no trace in the genome, aside from the modified gene. Complete metabolic pathways can thus be imported into living organisms, which become programmable mini-factories. We thus speak of synthetic biology. This new community has its own Mecca, the annual iGEM competition, where "the best genetically modified machines" are celebrated.[188]

As for any technological advancement, the prospects are both exciting and alarming. For instance, the PlastiCure-BGU project uses genetically modified bacteria to digest a common packaging plastic, PET (polyethylene terephthalate), resulting in a biological solution to the accumulation of plastics in the oceans. The technological prowess and good intentions of engineers are commendable, but could this also justify the pursuit of the mass production of plastic on Earth, or at least delay the necessary transition? Furthermore, synthetic biology companies would inevitably compete with traditional producers, often located in developing countries. The start-up Ginkgo Bioworks is already producing rose fragrance in yeasts. With synthetic biology, are we really reconnecting with the living being, or is it rather a form of hyper-optimization?

It should be noted that the production of GMOs is not negative in principle. Like any technology, one can get the best out of it or the worst. For example, if your child were suffering from a genetic illness associated with a growth hormone deficiency, you, like any parent, would certainly prefer that he or she were treated with a hormonal complement from a GMO (as is currently the case, and approved by the US

Food and Drug Administration since 1985) rather than hypophyses from corpses (as was the case prior to the transgenesis revolution). In spite of purification efforts, there is no guarantee that a hormone from a cadaver will not also carry certain diseases, such as Creutzfeldt-Jakob disease, whereas the transgenesis approach is much more controlled in this regard. When confined to a well-defined niche, optimisation can have its virtues.

However, in order not to give a blank cheque to the all-GMO, even in this biomedical niche, let us take the counter-example of the anti-malaria drug ACT (artemisinin-based combination therapy). When this molecule is synthesised by means of a GMO-type strategy, it is less effective than the natural molecule from the milled leaves of the *Artemisia annua* plant. Why would two identical molecules have different effects? A tonic based on Artemisia leaf is much richer in molecular interactions, particularly thanks to the flavonoids and terpenes contained in the leaf, than compared to a pill containing only the active ACT molecule. In this case, the simplification associated with synthetic biology eradicates valuable synergies.[341]

More generally, the debate on plant GMOs underlines the complexity of our relationship with living beings. For anti-GMOs, one of the arguments often put forward is the sacrilege of modifying life. Yet most of our food comes from mutant plants selected over thousands of years of domestication by humans. The introduction of transgenes is also a natural process, for example in galls. It is rather an anthropocentric or even animal-centric view that prevents us from understanding that the plant kingdom allows much more genetic transfer and mosaics than animals.[342] The financial and economic argument of the anti-GMOs is certainly more relevant. Indeed, the pro-GMOs often stress the need for agricultural yields to ensure food security, even though GMO-based agriculture is currently leading to a worldwide simplification of agricultural production, to the subjugation of farmers to the large seed companies, and thus to a weakening of our agricultural systems.

Finally, with circular economy, bio-economy and synthetic biology, have we really stepped away from the contradictions of sustainable development? While technologies make a great deal possible, by default, our economic and psychological biases further commit us to optimisation. Turnkey and reductionist solutions are very attractive. It should be noted that this trajectory is widely shared, even among the greenest of ecologists. For instance, some vegans are the leading advocates of synthetic biology, under the guise of the preservation of animal welfare, without necessarily considering other negative externalities. Undoubtedly, optimisation is a simple solution that is easy to mobilise in all sectors.

6.3 ANOTHER INFINITY

We have seen how optimisation has been a powerful driver of the Anthropocene and its consequences. In response to this, the promises of the bio-inspired world of tomorrow open up genuine paths to the future, but also present the risk of a form of total optimisation, perfectly compatible with the ideology of the *good Anthropocene*. We find ourselves halfway through our reconnection to Earth, while still being at risk of going down the wrong path. Given the many past mistakes, can suboptimality provide a viable and systemic alternative?

The examples provided by biology show us that the factors that are perceived negatively within the optimalist model are in fact robustness factors. Suboptimality builds robustness from the vulnerabilities of living systems. All instances of incoherence, slowness or redundancy inhibit performance, but also enable adaptability. We have also seen how life is built on contradictory interactions, between tension and compression, or between randomness and redundancy, with a view to establishing autonomy and robustness. Finally, suboptimality considers the value of emergent properties of interactions. It does not hide the associated costs; on the contrary, it embraces their counterintuitive effects to create a new value: robustness. As such, it opens up a new infinity. This is where such a third way can be engaging.

Favouring adaptability and transformability over control and optimisation means acknowledging that we have left the Holocene and its astonishing climate stability behind. With robustness, we would then be responding pragmatically to the dominant character of our present and future times: social and environmental fluctuations. This new trajectory of progress is very open-ended, and far more exciting than human development based solely on performance increments. Such a destination could mobilise a large collective. In contrast to sustainable development and its many contradictions, or degrowth and its negative tone, robustness thrives on a deep human drive towards lasting and transmitting to future generations. Before exploring its properties and its pioneers, the link between infinity and personal commitment deserves a short digression.

Infinity has always been a mobilising force for human societies. A first infinity was, and still is, the divine infinity: the promise of life after death. In recent history, the materialistic infinity has progressively imposed itself over the divine one. Materialism is a question of flourishing on Earth, that is using natural resources. It should be noted that the conflict between these two infinities is not new and is already documented in antiquity. We have seen, with Georges Bataille, how Protestantism was an important step towards the material infinity in our western modernity.[200] By putting the compass on the earthly returns of human activities, Protestants created one of the greatest paradoxes of the Anthropocene: a disconnect to the physical world, created out of a desire to reconnect with Earth. A prior contribution to the questioning of the divine infinity can be added here: the discovery by Christopher Columbus of a new world absent from the divine scriptures calls into question the Bible and promotes the scientific school of thought. New universities turned away from the then prevalent theology and gradually established the scientific disciplinary fields that we are familiar with today. Materialist infinity, combining performance and growth, would be, at least in part, pragmatic responses to the limitations of the divine infinity.

Today, planetary boundaries are the death knell of material infinity. That infinity is finite. What other infinity could mobilise us? It looks like the 21st century will be based on the *infinity of interactions*. Consider three contemporary societal revolutions that are all revolutions of the infinity of interactions: the feminist revolution that attempts to reconnect the two halves of humanity, the ecological revolution that attempts to reconnect humans and non-humans and the digital revolution whose main object is intermediation. These three revolutions do not simply demand more justice or more equity for certain communities. They create new properties and new values for society as a whole. In the examples on the following pages, we will also see how cooperative strategies emerge when scarcity becomes apparent, whether in biological or social systems. The infinity of interactions could thus emerge as a reaction to the limits of materialist infinity.

The infinity of interactions also resonates with the divine infinity. They come together in their topological essence. God is sometimes considered as a social object, emerging from the human faculty for communication.[343] The word *religion* comes from the Latin word *religare* (*to link* or *to bind*). The divine infinity, if we consider God as an emergent property of human interactions, would thus belong to the infinity of interactions. In this staggering scenario, the infinite nature of the relations between humans and non-humans is indeed divine "by nature": the digression of the Anthropocene, and its materialist infinity, has transiently disconnected us from this well of abundance, instead leaving us to exhaust the resources of the planet. Our answer, the infinity of interactions, would therefore be a first step towards a critique of optimisation and its simplicity, and a more intimate voice allowing us to understand our place and our connections within this world.

More operationally, and thanks to the emergent and amplifying properties of cooperation, the infinity of interactions can create new economic and social values without exhausting material resources. As we reach planetary boundaries, our future path necessarily lies somewhere in this infinity. It is time to explore which form this robust world might assume, beginning with the foundation of any sustainable society: food resources.

6.4 AGROECOLOGY

If we wish to invent a truly bio-inspired world, removed from the tropism of performance, and capable of providing the majority of our resources without any negative impact on ecosystemic services, then we must give very serious consideration to the option of agroecology.[55]

Agroecology uses slower and more heterogeneous methods. It perfectly illustrates the suboptimal society under construction.[55, 186] For instance, rather than using the solutions of the so-called green revolution, which stimulate short-term returns but cause desertification in the long term, agroecology mobilises the natural diversity and cycles of ecosystems, in order to maintain sustainable agrosystems. Olivier de Schutter, defines it as "the fusion of two scientific disciplines, agronomy and ecology" and as a set of practices, the objective of which is to "improve agricultural systems by imitating natural processes, thus creating beneficial biological interactions and synergies between the components of the agroecosystems".[344] The Afterres2050 scenario produced by the Solagro research organisation emphasises the feasibility of such an alternative for human food, provided that livestock farming and the consumption of dairy products and meat were significantly reduced.[55]

In the field, the use of external inputs and pesticides is reduced. Local resources, including energy resources, are drawn on as a matter of priority, in a form of circular agriculture. How can one compensate for the absence of these services? Agroecology finds them in the well of abundance of interactions: in the soil, between the cultivated species or even with livestock. How is this possible?

Agroecology prescribes a reduction in the working of the soil. In fact, while labour remobilises the mineral elements of the soil, it also causes its mineralisation and therefore its desertification. In order to remain alive, a soil must be rich in organic materials. Therefore, it is not a case of intermittently increasing the quantity of nutrients using synthetic fertilisers, but rather of giving the soil the means by which to maintain its richness in the long term autonomously. Once again, robustness replaces performance. This occurs in particular through permaculture, which implies a continual covering of the soils with plants. The ploughing of large farms in so-called conventional agriculture, exposing the land to bare soil for several months of the year is simply aberrant.

More fundamentally, agroecology emphasises the benefits associated with biodiversity. For example, let us consider the classic example of the co-cultures that combine maize, beans and squashes. This "divine trilogy" was practised by the Mayans, and has now re-emerged under the term "milpa" in Mexico. What are the mutual benefits of such a practice? Beans, leguminous plants, have nodes with symbiotic bacteria in their roots, which enable the fixing of the nitrogen in the air: they enrich the soil without the need to add external nitrate. In return, the climbing bean uses the maize as support. The large leaves of the squashes on the soil prevent the development of weeds and create a favourable humid microclimate, which limits the need for irrigation. Finally, the quantity of total biomass per square metre is excellent, since three species live alongside one another on the same parcel, within a temporal succession, wherever a large cereal farming operation has an excellent grain yield, but only during a small period of the year.

Co-cultures can also concern a more restricted range of species, while retaining several advantages. Thus, in reaction to the single-crop cultivation of cereal promoted by agroindustry, varietal mixtures of cereals are becoming more and more frequent in fields. Rather than targeting optimal performance by sowing the best possible species on a parcel, different varieties are instead sown together, along with their divergent strengths and weaknesses, while still ensuring that the harvest can be more or less synchronous. Owing to environmental hazards, the type of grains to be harvested is not guaranteed. One variety could suffer more than another. However, in contrast to monoculture, which runs the risk of total loss, for example, in the case of disease, varietal mixing supports yield stability. The harvested grains are heterogeneous, and it is this heterogeneity that enables the robustness of the crop. Moreover, studies show that, during their growth, varietal mixtures develop synergies and cooperations that allow the whole field to better resist drought and pathogens.[345, 346] Because of the advantages described earlier, varietal mixtures are increasingly used by farmers. Between 2017 and 2019, varietal mixtures increased from 5% to 12% of the cultivated wheat area in France. In some French departments, a true shift has even taken place, for example, in Mayenne (near Brittany), with the rate rising from 1% in 2009 to 40% in 2020.

This varietal mixture approach also presents operational challenges, fuelling many promising scientific projects.[345] For instance, research in agronomy has long tried to identify the factors that control organ size, with the underlying idea of increasing yields. This was the age of performance. Today, with the shift towards robustness, we are rather addressing the question of the stability of yields in a fluctuating environment. This is the age of robustness. It is safe to say that we have only brushed upon the vast array of possibilities presented by suboptimality in agroecology. A number of counterintuitive results have already been obtained. For example, it has been shown that plants cultivated without phosphate produce excellent yields, provided the soil is equally deficient in iron. It appears that the network of interaction within the mineral elements of the soil, their transport and their metabolism, opens up new paths by which to adapt to the scarcity of resources.[347] Likewise, and this time as the result of participatory research conducted by farmers themselves, vegetable crops without any irrigation have managed to produce astonishing yields.[348]

Another form of co-cultivation, agroforestry – that is, the combination of trees and vegetable crops, vines or field crops – is also attracting increasing attention. In particular, in a country such as France, described by Julius Caesar as *hairy* on account of its many forests, the savannah-like monocrops of the Beauce region near Paris appear thoroughly irrational. Throughout the course of agricultural history, we would have copied a model inspired by the Fertile Crescent, without asking ourselves whether it is suited to our temperate climates. Trees provide several benefits within agrosystems, such as the stimulation of soil life thanks to several symbioses. The cover of the canopy also maintains a microclimate that limits the need for irrigation, and provides a buffer when faced with climatic variations. Finally, agroforestry is not incompatible with a certain degree of mechanisation, provided the trees are sufficiently spaced. Research involving market gardeners and farmers not only demonstrates the feasibility but also the profitability of this new model of more integrated agrosystem.[349]

Having a sound knowledge of ecosystems also enables the restoration of desertified sites. In particular, within a typically suboptimal scenario, often-despised weeds are increasingly seen as pioneer plants capable of restoring soils that have been degraded by deforestation or intensive agriculture. Remarkable results have been obtained in Australian territories that were formerly desert.[350] Such an approach is also in line with that of landscape architect Gilles Clément, who condemns the use of aggressive, warlike techniques. Thanks to the properties of local weeds, "the garden authorizes disarmament".[351] Even deadwood in rivers gains letters of nobility, thanks to its systemic impacts. In particular, a better understanding of its ecological role inspires gentler approaches to river maintenance. It is a question of "working more with the river, not against it".[352]

Finally, the question of enhanced genetic diversity goes beyond plants and soils and also includes livestock, which can be raised on the same parcels. Livestock farming indeed benefits from this agroecological revolution, with initiatives that were unthinkable just a few decades ago. For example, aromatherapy based on essential oils is now being used to replace antibiotics, with increasing frequency and to great success.[353]

Finally, agroecology is opposed to precision agriculture, which relies on the fine mapping of territories to optimise the sowing, maintenance and watering of each plant. This form of agriculture makes extensive use of digital technologies, prioritises control and optimisation, promotes monocrops and grants excessive power over natural resources to just a few digital companies. In 2006, two Google employees founded The Climate Corporation, a major player in precision agriculture in the United States. In 2013, this company was sold for $1 billion to Monsanto. For some, the acquisition of Monsanto by Bayer in 2016 (for $66 billion) was not only due to the company's biotechnology portfolio but also due to its digital maturity.[354] Faced with this model of total control, the suboptimal counter-model could claim a form of "imprecision agriculture", where decision-making is decentralised, heterogeneous and even incoherent, where the final products are not uniform, where some waste of arable or resource-rich land feeds into ecosystem services, where technologies – chemical, digital and biotechnological – are optional (another guarantee of robustness to social, geopolitical and environmental fluctuations). Agroecology defends a diversified and situated agriculture led by technically autonomous peasants (etymologically, from old French, "pais": country). It is opposed to a form of extraterrestrial agriculture, where control is delegated to technocrats and operations are assigned to farmers who are reduced to the status of subcontractors.

Initially thought of as a new, sustainable agricultural system, agroecology requires more significant mutations. In fact, in order to be genuinely functional, it also requires the regulation of the food markets to guarantee a fair salary for peasants, mutual financing systems, committed politicians and policies, the training of producers, the education of consumers and citizen science. Moreover, the development of Community Supported Agriculture (CSA) demonstrates that consumers are increasingly engaging with this issue. In Japan, the Teïkeï, a local form of CSA, impact almost 3 million consumers.[55] Agroecology thus amounts to a profound agricultural, economic, political and social transformation, which agronomist Michel Griffon incidentally refers to as the *doubly green revolution*.[355] The notion thus emerges that suboptimality is not merely a safeguard for the bio-inspired world of tomorrow but that it also raises the question as to the type of economic system that would facilitate the robustness of human societies.

6.5 FOR AN INCLUSIVE SOCIAL MODEL

For the well-being of the population, human societies have no choice but to harvest resources from their environment. There exists a lower threshold, beneath which an excessively poor harvest would result in unsustainable hardship. In the Anthropocene, as we have seen with the concept of planetary boundary, there also exists an upper threshold, above which human societies are not viable, however, in this instance for reasons of ecological feedback. Our economic system must therefore be contained within these two limits.

The term donut theory is occasionally used as a metaphor for humanity's area of viability: neither too few resources to enable survival nor too many resources to enable coexistence with the planet.[356] In a way, this is a new metaphor for Aristotle's happy medium.[357]

This compromise is a further appeal for suboptimality: while technologically speaking we can continue to expand and improve our performances, we should instead learn to accept, use and live with the constraints of our environment, just like every other living creature on Earth. For example, in the case of circular bioeconomy, it would not be a question of optimising the whole landscape, but on the contrary of acknowledging that not all of the landscape can be optimised. This would amount to sanctuarising certain territories, with no obvious economic development in the short term.

In which socio-economic system can this happy medium be achieved? Extremist systems are excluded, that is sectarian, dictatorial or populist systems, which aim at a high degree of synchronisation and homogenisation, and are therefore self-defeating in the long run. I shall now focus on the two main economic visions within the western world and where heterogeneity and creativity can be expressed.

Let us begin with the liberal model,[358] founded on entrepreneurial freedom and the limitation of regulations. The individual lies at the heart of this model. As Margaret Thatcher summed up, in a liberal system, "there is no such thing as society". The liberal model, in essence, is a great support for individual creativity and could contribute to a healthy economic heterogeneity. But this individualistic model also actively promotes competition. As in biological systems, this can lead to a form of homogenisation, and thus to the opposite result: the most competitive can crush the others or prevent them from emerging.

Furthermore, individual freedom promotes individual comfort, and is therefore, by construction, a blank cheque for our individual cognitive biases. As we have seen, the idea of infinite economic growth, made possible through an unbridled economic liberalism, is also the individual symptom of a self-justifying cascade of engagement. The famous invisible hand (of the market) is rather an invisible cognitive bias.

If the risks associated with cognitive bias are well known, is there a quantitative demonstration of the deleterious effects of such an individualistic drift? The analysis of social networks could provide important answers. The emergence of "fake news" is indeed a manifestation of the increasingly central positioning of the individual. By converting opinions into information without any delay or filter, we synchronise ourselves with our individual, and therefore biased, world view. For example, it has been shown that false information has a 70% greater chance of being republished than true information on Twitter.[359] A random decision, without cognitive bias, would be more scrupulous! A safeguard mechanism against individualistic drift is urgently needed. In this respect, the excesses of the liberal model are perhaps the best advocates of the social model.

The social model is organised on a supra-individual level, that of human society. In contrast to economic liberalism, it hampers our individual cognitive biases. It favours the robustness of the group to the detriment of individual comfort. In this sense, this social model shares a common objective with suboptimality. In other words, the social model does not promote heterogeneity by liberating individual actions. It provides a framework within which heterogeneity can and should be expressed. It also offers an operational solution to our individual psychological limitations, by putting the focus on the group. A philosophy also embodied in the southern African term *Ubuntu*, and its proverb: "Umuntu ngumuntu ngabantu" (I am because we are). Compared to the liberal model, the social model may seem slower and less responsive. One might say that it is less efficient. But it responds to the rebound effects of the liberal model, which, by displaying unwavering support for individual creativity, tends instead to create homogeneous systems guided only by cognitive biases.

In recent western history, these two models have been dominant alternately. In particular, the social model was prevalent following the economic crisis of 1929 and in the post-war period, that is in a world facing scarcity.[360] Its expression can be found in the Havana Charter of 1948 mentioned earlier: this charter accepted the importance of a fair price controlled by supply and demand, but it went far beyond by also recognising the shortcomings of a naive "belief" in the market economy: preservation of natural resources, food security, reduction of inequalities between poor and rich countries. In short, it aimed at regulating an over-simplistic free trade. The operational products of the Havana Charter, the General Agreement on Tariffs and Trade (GATT) and then the World Trade Organization (WTO), omitted these essential subtleties in a world of economic growth and opting for the liberal model.[125]

Would the liberal and social models be economic responses to the perception of a world of abundance or of scarcity? Let us embark on a biological detour and consider one of the great laws of ecosystems.

6.6 TOWARDS A WORLD OF COOPERATION

In living systems, abundance of resources tends to stimulate competition. For example, in the case of the algal blooms mentioned earlier, nutrient affluence coupled with favourable environmental conditions stimulates the proliferation of cyanobacteria colonies. In turn, this homogeneous phytoplankton occupies the available space rapidly and even produces toxins to kill competitors. Similarly, in soils rich in organic matter, sexual reproduction is almost absent: most organisms reproduce clonally and therefore compete for the same resources.[295]

This law can be found in human societies: economic growth supports the values of competitiveness in a society of apparent material abundance. In this framework, the liberal model would simply be an economic variation of a biological law: abundance of resources allows and stimulates competition between individuals. The tropism towards performance and individual freedom is therefore well adapted when environmental conditions are favourable. This is the way of materialistic infinity. However, this individualistic path is not sustainable in the long term, not only because the environment is always changing but also because strong individual performance accelerates the tempo of these changes and leads even more quickly to scarcity. Put differently, from a biological point of view, the liberal model is the winning card of the individual and the (very) short term in a world of material abundance.

On the contrary, scarcity in ecosystems stimulates cooperation: value is created not through material resources, but through the heterogeneity of partners. This is the way of the infinity of interactions. This behaviour is the basis of all biological symbioses. To be convinced of this, you can consider a forest. When environmental conditions are favourable, if you remove a tree, the neighbouring trees will take advantage of the clearing to increase their biomass. This shows that trees were competing against each other before. Again, the abundance of resources stimulates competition. On the contrary, if you carry out the same clearing in much harsher environmental conditions, such as in high altitude, the trees neighbouring the removed one will reduce their biomass. This shows that they were cooperating before: their biomass decreases because they just lost a partner.[361] This law can be verified in many biological contexts. For example, penguins in the Antarctic blizzard group together and perform a form of social thermoregulation: some individuals move to the periphery of the group in turn, and the individuals in the centre of the colony are protected from cold. Similarly, yeast survive metabolic shortages when co-cultured in a test tube because their deficiencies are compensated for by sharing their metabolites.[362] This is the true power of social life: it allows the group, and therefore the individuals, to respond to environmental hazards with expanded adaptive capacities. Therefore, from a biological point of view, the social model is the winning card of the group and of the long term in a world of scarce resources.

Economic liberalism can, all the same, provide support for individual initiative, and thus for creativity, which strong traditions or a sclerosing economic system would prevent. Historically, the liberal model was indeed imposed in response to the stranglehold and to the threats of the Stalinist centralised model during the Cold War. As is often the case, it is more a question of balance. Some liberal-inspired performance could be justified if it were limited in time or to a niche. The problem today is the structural and ideological domination of the liberal model over the social one: some local and short-term performance indicators are improving, but our overall robustness is weakening. This debate between local and global, short term and long term, is a key point in the conflict between the social and liberal models. For example, supporters of economic liberalism often emphasise the socio-economic gains for otherwise underdeveloped countries. However, it is all a matter of time. I am not convinced that the residents of the Maldives still see their development, and that of their neighbours, as a genuine benefit, their territory being condemned in the medium term by the rising water levels and the climate crisis.

One might therefore be surprised by the permanence, and the domination, that the liberal model enjoys today, in spite of the fact that the environmental issue is becoming an existential question. As winner of the Nobel Prize for Economics Joseph Stiglitz put it, "neoliberalism must be pronounced dead and buried".[363] In fact, it can be explained quite easily: as we have seen with planetary boundaries, we live in a world of ecological debt. In other words, we are objectively in a state of resource scarcity. But we have invented an economic system that makes us believe in a world of abundance. Thus, we continue to emphasise the values of competition, further aggravating resource scarcity.

Knowledge of biological systems brings good news: when scarcity becomes inevitable, we will most likely favour cooperation. The multiple revolutions of the infinity of interactions – ecological, feminist, digital – are all weak signals of this world of cooperation in the making. Therefore, in order to anticipate this society, reducing our environmental impacts, as an act of ecological charity, is already outdated. Rather, it is a matter of considering that we are already in a state of systemic resource scarcity, without waiting for the economic system to declare it. Cooperation will follow.

6.7 SOCIAL DILEMMAS

Let's go one step further. Why would a model based on cooperation be more robust than one based on competition? What would be the consequences for individuals and their comfort? By nature, in a world built on individual freedom and competition, the most competitive win. This model is therefore perfectly compatible with the pursuit of the improvement of performance. But for how long? Competitiveness feeds an amplification loop. It therefore contains its own trajectory: in essence, competitiveness is the prelude and then the anthem to violence, first economic and then military.

It would therefore be more accurate to say that, in a world built on competitiveness, the most violent win. This, incidentally, is one of the scenarios emphasised in the report by the Central Intelligence Agency (CIA) for the world of 2035.[364] The most fervent supporters of economic liberalism often recognise the limitations of that ideology: laws generally prevent human societies from getting there. Even the CIA imagines, in a variant of its belligerent scenario, an ultimate return to reason, with a view to promoting cooperation and avoiding a global apocalypse. Again, the invisible hand of the market is not particularly robust.

Finally, the social model, comprising the virtues of an imperfect global regulation, is a necessary condition to the expression of suboptimality within the economy. This model is capable of incorporating planetary boundaries through regulations, whereas the liberal model would circumvent them by prioritising maximum, rapid human development, at any cost. The social model offers a supra-individual solution to the challenges of the Anthropocene, whereas the liberal, optimalist model places its trust in the creativity of individuals. The opposition presented above between imprecision agriculture (agroecology building its robustness on interactions) and precision agriculture (digital agro-industry building its performance on optimisation) follows the same logic.

Here we come to the concept of the social trap, which is derived from game theory. When individuals promote individual comfort, they can endanger the entire group, and therefore themselves, without knowing it.[365] Conversely, cooperative behaviour within the group can facilitate its survival, and even allows a certain degree of individualistic, selfish behaviour within the population. The social dilemma in the Anthropocene poses the question as to how much individualism we can still afford. Very concretely, this amounts to questioning the merits of our consumer choices – buying a car, using air conditioning, eating meat, etc. – which are all social traps. The Covid-19 pandemic provides a taste of the global social dilemmas to come, or how the discipline of each individual becomes necessary to preserve the population.

Are humans more liberal or more social? To those who would like to use Darwin's books to justify a liberal system founded on individual freedom and competition, let us recall that the theory of evolution states that selected individuals exhibit satisfactory traits, no more and no less. In fact, according to Darwin himself, and based on the past evolution of the human line, the first humans appear to have been removed from the principle of individual natural selection, prioritising social behaviours.[366] Indeed, and contrary to the still widespread dogma of an original human society guided by "war of all against all",[367] archaeological data instead suggests the existence of altruistic behaviour in the societies of the Upper Palaeolithic. For example, the bones that we have recovered thus far rarely show signs of violence. The human societies of the Upper Palaeolithic were very probably peaceful and egalitarian.[8] It is unlikely that the first humans favoured the natural selection of the most high-performing individuals. Rather, it is believed that they developed and evolved social capacities as a means of sustaining the robustness of the populations. Again, this could be a coping strategy for the shortage of resources. Natural selection would thus mainly operate at the group level, not at the individual level. The first human societies were social and suboptimal.

How can we explain the social condition of humans? Bipedalism could be an important key. On the one hand, it enables the opening of the thoracic cage and thus oral communication, a vital prerequisite to complex social capacity.[343] On the other hand, it brings the primate down from the tree and places the human firmly on the ground, that is, within a demarcated territory and under threat from predators. The social behaviour of human beings could thus represent a response to this new gift.[8] Today, the Anthropocene places us literally "on Earth": the finitude of the world has become our greatest predator. It will thus be necessary to discern a third path, probably social and inclusive, that is, an approach that envelops the rest of the biosphere, with a view to coexisting with the other inhabitants of the planet in the future: a form of *coviability*.[368] This implies new rights and duties.

6.8 THE SOFT LAW OF NON-HUMANS

If a social model, expanded to incorporate the biosphere, should come to fruition, it must also involve new laws. We must assign a legal and social status to our natural environment, no more and no less. Heresy? This is rather a new step in an irremediable historical trajectory. As Michel Serres says in the preface to *The Natural Contract*, humans already have conferred more rights to objects in the past. In particular, when looking at the terms of marriage along history, one can only be horrified when reading that women were also considered as objects, belonging to men, for the longest time. Women have now the same rights as men in a majority of countries, at least in theory. Although this is certainly too slow, gender equality is high on the UN agenda. The feminist revolution is a sign of the shift in society from performance to robustness, and it signals the failure of men to control women. In parallel, the ecological revolution signals the failure of humans to control nature. Finally, when ecofeminists denounce the devastating effects of patriarchy on society and ecosystems, the voice of a multifaceted exclusion is heard. Thus, the ecofeminist call is not reduced to a gender issue, it invites and invents a new relationship between living beings around a common cause: stop the tyranny of performance!

At least, the feminist revolution shows that alleged objects could indeed be turned into subjects during history. In such a trajectory, how could one give rights to non-humans today? In the words of Cicero, "we are slaves of the law so that we may be able to be free". The extension of rights to non-humans in the Anthropocene may invite us to revisit the very notion of freedom. According to Article 4 of the Declaration of the Rights of Man and of the Citizen of 1789, "liberty consists in the freedom to do everything which injures no one else: hence the exercise of the natural rights of each man has no limits except those which assure to the other members of the society the enjoyment of the same rights. These limits can only be determined by law". Thus, the definition of freedom is negative. It should not be maximal, and must therefore be restricted in order to guarantee the robustness of societies. In order to overcome the shortcomings of the Anthropocene, not a single word of this definition should be altered, except that the scope of the term "another" should be expanded to include non-humans and future generations. This would simply amount to correcting the economic aberration that considers nature to be a free resource, disconnected from humanity.

The redefinition of freedom in the Anthropocene, considering the term "others" in a broader, biospheric sense, could help us to overcome the opposition between the liberal and social model. In proposing to give a voice and rights to nature, *The Natural*

Contract without doubt provides us with a solid track for its implementation.[1] Michel Serres actually invites us to envision mechanisms enabling the mobilisation of new responsibilities for humans, within a natural environment that has become cultural and unpredictable in the Anthropocene. Thirty years on from the publication of *The Natural Contract*, we are beginning to see these mechanisms emerging, especially within the legal field.

In 2008, Ecuador was the first country in the world to give nature a legal status, under Article 10 of its constitution. This was about giving rights to Mother Earth, *Pachamama*, such as the right to exist, to regenerate and to maintain its structure and its functions. "Another" can therefore indeed be non-human, from a legal perspective, with obvious consequences regarding the contours of the individual freedom of human beings. Conversely, under Articles 71 to 74 of this same constitution, citizens can call on the government authorities to enforce adherence to Article 10.[108] The movement appears to be under way. Other states such as Bolivia have followed Ecuador. The Ganges acquired legal status in 2017, although in this case Hindu nationalism likely played a significant role. Lake Erie in the United States or the Loire in France are likewise in the process of acquiring this right.[369] Perhaps most tangibly in our everyday life, the issue of animal welfare is increasingly being debated, and thus supports the cause of animal rights.

Echoing suboptimality in its wording, these new rights for non-humans fall under what is referred to as *soft law*, that is, a law whereby the legality is flexible, non-mandatory and open to debate. Thus, a law that is inefficient. Having first emerged in international law, in particular through the United Nations, and then having been transposed onto environmental law, the emergence of soft law constitutes another weak sign of our trajectory towards suboptimality. Indeed, soft law does not rely on performance in order to be effective. On the contrary, it capitalises on its relative weakness in order to access entities, the size or function of which is beyond our control. Being non-binding in nature, the strength of soft law resides rather in the dynamic that it generates, and thus in the actions that it legitimises. From this perspective, this is without doubt a strength: a hard law would prevent widespread compliance, for economic reasons if nothing else, and would be nipped in the bud.

We must, however, remain cautious. Giving rights to the living could be another form of disarming idealisation of life. It is not a matter of "whining about living things".[370] Declaring solidarity with mistreated non-humans, without questioning our economic system in depth, would be sterile. To choose to grant rights in the first place may obscure the duty and responsibility of each individual. What tool could complement the rights of non-humans?

If the Anthropocene invites us to extend the social contract to future generations and to non-humans, should we not first of all acknowledge our debts? This is what Léon Bourgeois proposed in his seminal essay "Solidarity" in 1896:

> It is this idea of man's debt to other men which, giving in reality and in morality the foundation of social duty, gives at the same time to liberty, to individual right, its true character, and thereby its limits and guarantees [. . .] It is a question of men, as joint partners, recognising the extent of the debt that each one contracts towards all by the exchange of services, by the increase in personal profits, in activity, in life resulting for

> each one from the state of society; once this burden has been measured, recognised as natural and legitimate, man remains truly free, free of all his freedom, since he remains invested with all his rights.

This view, which gives primacy to debts over rights, deserves to be revisited today. In particular, this very constructive vision of debt allows us to consider humans in an intergenerational trajectory. It places usufruct before ownership, interaction before predation. Putting debt before rights might therefore also accompany the shift from the condemned materialistic infinity to the infinity of interactions and its many emerging properties. Finally, non-human rights and human debts come together in the formalisation of a new relationship to our environment, to our resources and between humans. More operationally, the "commons" may well embody this new paradigm.

6.9 THE COMMONS

A resource is viable when it can feed a flux, while safeguarding the stock. According to the thesis of American ecologist Garrett Hardin in his article "The tragedy of the commons", all use of shared resources would result in their disappearance, in a typical social trap.[371] This threat of the commons is often cited by advocates of neoliberalism. Here the retort will be that it is rather the simplistic optimalist vision of nature that cannot envision any path other than that culminating in collapse. Suboptimality, taking into account the complexity of interactions, could indeed open up a sustainable path. As environmental law theorist François Ost puts it, the hypothesis of the tragedy of the commons is quickly dismantled once it is understood that a common good not only requires a shared sense of belonging but also a community-based method of management: "a collective doing", and not just an individual or shared "having".[108] But how can this "collective doing" be rendered effective?

Elinor Ostrom, the first woman to be awarded the Nobel Memorial Prize in economy in 2009, made a seminal contribution to the issue of governance of the commons. She observed that centralised solutions, whether by the private or public sector, did not allow the sustainable management of common resources. A third, robust path assumes a much more distributed form.[372] Following in the footsteps of Darwin who highlighted the cooperative abilities of humans, yet departing from the principle of competition and individual selection,[366] Ostrom conducted an advanced meta-analysis of cases of the self-organised management of resources. She chose robust examples, in some cases spanning several hundred years, such as forest and pastoral mountain parcels in Switzerland and Japan, or irrigation canals in Spain and Philippines. She deduced eight fundamental principles, all essential to the sustainability of the commons:

- define the framework of the common good to be managed (limited access),
- adapt the operating rules to the resource to be managed,
- allow the rules to be modified through a participatory approach,
- assure users that the entity overseeing the management of resources is also accountable to them,

- put in place (initially very weak) sanctions in the event of non-compliance with the rules (sanctions must primarily be a reminder of obligations),
- allow rapid and inexpensive access to a local dispute resolution body (in particular because any rule retains some ambiguity),
- be recognised by external entities as a self-organised structure,
- set up a modular organisation by function.

The ingredients of suboptimality are everywhere. In particular, a key point among these principles is the gradation of sanctions: sanctions that are very much symbolic are consistent with an almost voluntary compliance with the rules, an essential prerequisite to the sustainable management of resources. It comes down to favouring group education before individual sanctions. Counterintuitively, the "efficient" solution of strict and strong sanctions would lead to the exhaustion of the resource in the long run. Not only do very weak initial sanctions maintain the resource over the long time, but the resulting robustness accommodates a share of fraud over time. The robustness of the governance of the commons emerges from well-understood weaknesses.

Cooperative resource management calls for a reversal, echoing the current shift from the liberal economic model to the social model. We can already see that the value of individual freedom is being undermined in favour of the values of health or security. However, we are still at a crossroads. If this shift remains trapped in the optimalistic dogma, then we will favour the tracing of individuals or over-investment in personal protection and self-defence. If we shift to a model that emphasises group robustness, we will build common health or global food security. Again, the most robust trajectory is easy to identify once it is understood that group robustness preserves individuals, and on the contrary, that individual hyper performance threatens the group and therefore ultimately the individuals. With Elinor Ostrom, the (admittedly pastoral) management of commons offers at least one example of social robustness (integrated health and food security) built against individual performance. In the end, it is a matter of reappropriating the original definition of freedom as a product of social constraints.

However, the collective management of the commons requires effort and education. With the collective management of common goods, the idea also emerges that time will have to be freed up to make room for cooperation and the related pedagogy. The precedents of the "mutual school" and its more recent variations (e.g. in Sugata Mitra's self-organised learning environments[373]), may show us a promising way forward.[374] In these educational contexts, pedagogy is not top-down, but mainly horizontal: students who have understood teach others. In the end, it is a matter of considering that education is first and foremost an emergent property of interactions between students. Not only does this type of pedagogy have virtues in terms of the robustness of the knowledge acquired but the method also provides expertise in teamwork over time. Last, but not least, this method democratises and desacralises knowledge. A first step towards open science.

Where the efficient turnkey solution requires no education and turns citizens into subcontractors of the economic system, suboptimality requires us to slow down and to change the instructional goals: learning by the group for the group. Robust long-term social solutions will have to be preferred to short-term performance solutions. To go beyond words, let us explore the structural implications of such a reversal. Let us consider in particular the status of work in the post-Anthropocene society, and utilise certain ingredients of suboptimality in order to question it. Some ongoing developments are opening up the field of possibilities.

6.10 REDEFINING WORK

The thinkers of Social Solidarity Economy sometimes point to the prospect of universal income as a possible response to the challenges of the Anthropocene. What exactly would this entail?

There exist several variants of universal income. In its least restrictive version, as discussed here, each citizen would receive an income that could be combined with other income sources, unconditionally and without deliberation, whatever the citizen's age or resources, and for life, from the age of majority. This is about as far removed from optimisation as it gets.

The working poor suffer from a dual scarcity: shortage of resources and shortage of time. One of the primary motivations for universal income is to combat both. It is a case of counterbalancing the collateral effects of performance. In a Technology, Entertainment and Design (TED) Talk that is still prominent today,[375] journalist and historian Rutger Bregman clearly outlines the main arguments, which I shall pick up on below.

When time or money are lacking, cognitive biases are exacerbated: we have a tendency to make bad decisions, based on the short term. Try going into the supermarket with an empty stomach, and you will note that your basket will no doubt contain more highly processed foods, with too much sugar, fat or salt when you are hungry. Psychologists have considered this question, and have even been able to demonstrate that limited time or resources reduce our IQ and narrows our thinking, by focusing on the short term.[376] This is known as decision fatigue: when our brain runs out of fuel, it saves itself and takes shortcuts.[377] Not only do we make bad decisions but we also do not realise it.

For fans of performance, the acceleration of time calls for optimisation. If you are lacking time, avail yourself of even more efficient processes. For example, make use of online planners to better manage your time, or distribute free mobile phones in order to facilitate communication between employer and employee. If you lack resources to live, take advantage of the opportunities offered by the uberisation of the labour market to bypass social heritage and work around the clock. This is the strategy of the "augmented employee", as it were. As we have seen with the Jevons paradox or counter-productivity, these performance-based solutions can function in the short term, but only shift the threshold of urgency. Ultimately, their systemic effects worsen the situation.

In contrast, universal income offers a prospect that echoes suboptimality. In theory, in abolishing the dogma of *working to live*, it also does away with the urgency associated with the lack of resources, by guaranteeing a safety net in the long term. Beyond its non-optimised nature and its clear distancing from the dogma of performance, universal income could bring about other emerging properties. Let us explore them.

Universal income could modify our way of living and consuming. For example, poor people show the worst indicators in terms of the consumption of alcohol, tobacco and highly processed foods. If psychologists studying severe poverty are correct, this behaviour could be eradicated with a restored long-term perspective.[376] If poor people were to consume better, this would also have impacts on their health. Is this actually true? The case of the City of Dauphin in Canada is among those to have received the greatest media coverage, in particular on account of the fact that significant positive effects on education and health were achieved during the few years in which universal income was tested.[378]

Could universal income promote laziness? Let us remember that universal income is not a substitute for average income, but rather offers a basic income from which to live with dignity, no more and no less. The incentive to action would therefore still be present. In guaranteeing prospects in the long term, universal income could actually encourage risk-taking. Is this credible? In Namibia, a trial of universal income resulted in a significant increase in the number of small businesses.[379]

However, in stimulating entrepreneurship, would universal income serve as a successful ally to the market economy? It could instead eradicate questionable jobs. In fact, it may be a response to the questions that are increasingly asked of employees and management staff in their professional activities, and increasingly early on in their careers. Consequently, during his graduation ceremony in 2018, young engineer Clément Choisne gave a bold speech that received media coverage, in which he discussed his unease: "unable to see [himself] in the promise of a senior management role, an essential component of a capitalist system of overconsumption".[380] For similar reasons, agronomist Mathieu Dalmais refused to obtain his own engineering degree, while also unfolding the shortcomings of a one-sided formation focused on performance in popular lectures.[381] Calls for scientists to "change jobs",[382] to "de-innovate"[383] or even to "stop research"[384] are also becoming more insistent.

The environmental crisis is the existential question of the century, and it is therefore also one of the best sources of meaning available today. As such, universal income could well become a powerful catalyst for the ecological transition. More generally, with advances in automation and digital technology, jobs are disappearing. In contrast, the environmental question is requiring more work than ever before. Universal income could be a response to this new economic and environmental shift: labour value no longer lies in the (individual) job, but in the meaning attributed to (collective) activity.

As opposed to the market economy, universal income supports the collaborative, more local and adaptable economy thanks to networks of interactions, not subject to the logic of profitability. Universal income echoes suboptimality, by supporting group robustness mechanisms in place of individual performance. There still remain question marks, with regard, for example, to the effects on inflation or on the long-term developments in certain sectors. The subject is still in its experimentation phase. In 2019, the French movement for a base income listed 9 completed experiments, 12 ongoing experiments and 12 others in preparation or under investigation around the world.[385]

Too revolutionary? Not revolutionary enough, according to some! Universal income does remain subject to criticism, even among supporters of the collaborative economy. In the era of common goods, feelings are mixed: this basic income (around 1,000 euros per month) would indeed enable the liberation of certain labour market activities; however, it would still justify the preservation of traditional employment at the same time, with a view to topping up the income and financing universal income within a capitalist system. Sociologist and economist Bernard Friot envisions a much more radical alternative: the salary for life.[386] In contrast to universal income, the salary for life acknowledges that people are active throughout their entire lives. In this respect, they maintain a level of activity that would justify a salary until the point of death. The value of the salary for life, between 1,500 and 6,000 euros per month, would depend solely on a level of qualification, and would do away with capitalism altogether. This merits an explanation.

Within the market economy, and most notably in the calculation of the GDP, labour is considered a productive activity when it increases the value of the capital held by the owner of the work tool. For example, cleaning your home, for your family, is not considered to be productive work, whereas doing the very same cleaning for a company, in return for remuneration, is a recognised form of productive work. Yet work, as a physical human activity, does not have this social connotation: cleaning requires a certain amount of energy, whatever the context. Is it possible to correct this economic anomaly? This would involve fundamentally redefining work, for what it is: an activity, without any dependence regarding the owner of the respective capital. In fact, this would involve a more profound change in the economic system. In particular, anyone who works would also be the owner of his or her own work tool.

While this perspective appears utopian to some, it is nevertheless presently emerging within Social Solidarity Economy. Within this new economic paradigm, the company or association uses its profits not for lucrative purposes, but to serve a social and/or economic cause. The modus operandi is also more collective and democratic, generally with a more direct link to the local territory. Cooperatives, which build on collective ownership, form the basis of this new economy, also acting in coherence with the necessary mutual management of common goods. Is this alternative anecdotal? Since the first Michel-Marie Derrion cooperative grocery store opened in Lyon in 1835,[387] cooperative companies have multiplied in number in France. There were 23,000 of them in the country in 2016, generating a total turnover of 317 billion euros. One in three French citizens is now a member of a cooperative. For instance, one can become a cooperator of agricultural lands: thanks to the *Terre de liens* association and contributions from private individuals joining this cooperative, lands are being removed from the speculative market and are being made available to organic farmers, operating locally via short channels and following agroecology principles.[388]

With Social Solidarity Economy, universal income and salary for life, the current ecological and neoliberal impasse drives us to create new ways of doing society. We are completely reshaping the notions of activity and labour. Within these revolutionary projects, it is indeed robustness that is emphasised, and not performance. The redefinition of labour, of course, also raises questions surrounding financing, and consequently surrounding the associated mechanisms of redistribution. The foundation of the social model presents an opportunity to revisit taxation, considering the necessary interplay between individual comfort and global robustness, once again echoing a form of social suboptimality.

6.11 THE VIRTUES OF PROGRESSIVE TAXATION

A key point in the conflict between the liberal and social model in economics relates to the acceptable degree of redistribution. The liberal model tends to reduce the role of the state, and thus of taxation, while the social model is built on a form of assumed redistribution. Within an economic world dominated by neoliberal ideology, the degree of redistribution through taxation has fallen significantly. The higher marginal tax rate in the United States, that is, the rate at which the richest are taxed, was 37% in 2019. Today, any attempt to increase this rate is generally perceived as a confiscation. Yet in 1960, during the time of the dominant social model, this rate was 91% in the United States.

The current discrediting of progressive taxation appears aligned with the hyper-optimisation of our societies. Taxation crystallises the conflict between global robustness in the long term, which is dependent on high quality public services funded by the community of taxpayers, and individual comfort in the short term, achieved through the reduction of contributions and increase in purchasing power. From this perspective, taxation adjusts the degree of suboptimality within our societies.

What would be the optimalist alternative to taxation? In the seldom-imposed neoliberal world, the super rich often prefer to keep the control over their method of redistribution in the form of charitable organisations or foundations. Incidentally, philanthropy is also an excellent driver of defiscalisation and is therefore an additional means of fiscal optimisation. Because, fundamentally, the decision-making process adopted by charitable organisations is individually driven, depending on the identity of the person who provides the funds and on the chosen thematic area, philanthropy does not offer the safeguard provided by the systemic approach enabled by taxation. For instance, we have seen previously how foundations can have negative collateral impacts, even though the intentions are in some cases good. Once again, avoiding constraints through control and promoting individual comfort are not, generally speaking, robustness factors. In the interest of fairness, it should be noted that some foundations genuinely lay down a degree of their control. In spite of these exceptions, the shifting of the taxation of the super rich to a form of sponsorship is increasingly subjected to criticism. For example, when invited to the WEF in Davos, a sanctuary of neoliberalism, Rutger Bregman, vehemently condemned the "stupidity of philanthropy", saying he felt as though "at a firefighters' conference and no one is allowed to speak about water".[389] In this case, the elephant in the room is taxation and taxes.

In the spirit of the happy medium, a question arises: What degree of flexibility will be given to the progressive nature of taxation? A much more documented macroeconomic analysis can be found in *Capital and Ideology* by Thomas Piketty.[390]

In Russia, all taxpayers are taxed at a single rate of 13%. In this fixed and non-progressive form, taxation is obviously not an inequality reduction factor. Even capitalist countries such as the United States have declined to adopt such a socially unjust form of taxation. At the other extreme, and thanks to new digital technologies, we could imagine a hyper-personalised version of progressive taxation, reflecting the financial status of each citizen to the nearest cent or metadata on social networks. An optimised version of taxation, as it were. The complexity and opacity of these rules would probably render the collection of taxes less transparent and less acceptable, not to mention the profound, and therefore poorly defendable intrusion into the private sphere. So, is the progressive taxation that we are familiar with today already suboptimal?

The tax threshold depends on a number of factors (salary, marital status, the number of children etc.), however, it cannot be personalised to the point of perfect adherence to the characteristics of each taxpayer. The very concept of a tax scale implies the absence of further optimisation. The weakness of an imperfect, arbitrary scale becomes a strength, since the shared and simple global rules are transparent and therefore acceptable in a typically suboptimal scenario. The importance of clear rules in order to be able to comply voluntarily is also one of the key points raised by Elinor Ostrom in her principles for the governance of the commons.[372]

In France, taxation in the 19th century essentially served the financing of the preservation of order. Progressive taxation only emerged at the beginning of the First World War, with a view to improving incoming revenue streams and thus supporting the war effort. The mechanism was then retained in peacetime – a typical ratchet effect. In this case, this increase in performance was beneficial, since it also enabled the financing of education or intergenerational solidarity (pensions). Today, in comparison to the 1960s, taxation is significantly less progressive. This is a result of a political alignment with the neoliberal ideology on the one hand, and growing competition provided by fiscal optimisation on the other. Progressive taxation does exist today; however, the associated benefits are lessened by the optimalist ideology. Finally, it perfectly illustrates the difficult balance to be found between too much and too little performance and personalisation.

The downturn in the economic situation imposed by the Covid-19 pandemic and following crises could eventually favour the social system and its redistribution-fuelled robustness. Whether for the management of commons, Social Solidarity Economy, or taxation, these perspectives imply a form of political stability, at least on a local level. It is time to put the political systems to the test of their own robustness.

6.12 DEMOCRATIC ROBUSTNESS

By analogy to biological systems, in which political system does one find redundancy, slowness and incoherence? Some would say in all of them! So, in which political systems can we simultaneously find these values, judged as negative, and robustness? By performing robustness testing during each election, democratic and parliamentary systems have proven successful in this regard, whereas dictatorships portray the illusion of control, in some cases for quite some time, right up until the inevitable final collapse. Dictatorship can be both very efficient in the short term, and then very quickly descend into disaster. The continuous imperfection of democracy renders it adaptable. This echoes again Ashby's law of requisite variety[307]: the impermanence and internal dynamics of a (democratic) system enable it to cope with fluctuations.

Ultimately, the act of democracy emerges from a response to a social dilemma. It is about fostering the conditions for cooperation to preserve the group. In particular, freedom of speech in a democratic system maintains the feeling of belonging.

In turn, it promotes the cooperative behaviour of individuals.[391] Thus, freedom of speech is not just about sharing a wide array of information; the unhindered ability to communicate also generates trust between individuals and in the political system.

In fact, trust is one of the fundamental emergent properties of infinity of interactions. This can be seen by considering the interactions, and their effects, within the group. A leader is often required, and it has been shown that strong ties within a group lead to the appointment of this individual by way of an election. In contrast, a group made up of more distant and individualistic people will prefer a more authoritarian leader.[392] The dichotomy between a democratically elected, changing and adaptable leader and a self-proclaimed stable and "perfect" leader also emerges from our ability to interact.

In spite of certain incidents, the slow expansion of representative democracy in the 19th and especially in the 20th century unveils the recognised values of heterogeneity (i.e. diversity of opinion and conviction) within our current political systems. This is a slow, yet probably inexorable movement towards robustness. Here follows a selection of contrasting perspectives.

Democratic representation has gradually become more inclusive, most notably with women being granted the right to vote. However, it must be noted that we had to wait until 1918 for the right to vote to be granted to women in the United Kingdom and until 1920 in the United States. It will be little consolation to note that the Swiss canton of Appenzell Rhodes-Intérieures did not grant this right until 1991.

The use of a system of proportional representation in elections also indicates a movement towards a better representation of the diversity of opinions. This type of election was implemented for the first time in Belgium in 1899. Currently, the election of the members of the European Parliament is a proportional voting system: the lower threshold for the admission of parties is 5%. Proportional voting does, however, present notable disadvantages, such as the fragmentation of the electoral vote, which oblige the formation of often-confused coalitions, after the fact. We are still in an experimentation phase when it comes to suboptimality in politics.

Democracy also incorporates delays into its modus operandi. Far from being an obstacle to decision-making, these delays are a well-known safeguard. For example, a dictator within an efficient system has a greater chance of pushing the nuclear button than a political leader within a democracy subject to the opinions and delays of parliamentary chambers, commissions and other committees. Incidentally, you can see here how delays and redundancy (the "administrative millefeuille" with its numerous layers evocative of the popular pastry) can be a strength when it comes to adaptability. In contrast, it is when the number of intermediaries reduces, and thus when redundancy is low and delays short, that the system becomes fragile and vulnerable. It should, however, be noted that the "administrative millefeuille" can also reveal the optimal nature of an organisation. Why add additional layers of administration if not with a view to the improved overall planning of a task. The addition of intermediaries is also the main factor contributing to the establishment of a pyramid structure within organisations. A form of self-justifying administrative performance. Once again, it is all about finding the acceptable degree of suboptimality to support robustness first.

Finally, political systems are increasingly calling for the integration of a degree of randomness. This refers to "sortition", or "stochocracy", where political officials are randomly selected from a pool of candidates. Well known within the context of jury trials, the benefits of selecting jurors by means of a random draw have long been recognised: it enables almost every citizen to be positioned as a potential juror. In order for the citizens to accept the justice system, they must be closely involved in its function. Moreover, it should be noted that this is suboptimal for reasons other than randomness, since an imperfect heterogeneity is preferred over the performance of competent professionals. In an institutional context, the Citizen's Convention for Climate was established in France by appointing members by random draw.[393] Similar to the case with soft law, this is perhaps an indication that society understands, accepts and even adheres to the virtues of suboptimality as a means to challenge the inertia in tackling the major issues of our time.

Is this perspective somewhat too angelic? Our inability to see the counterintuitive benefits of diversity, delays or randomness in politics could pose a threat to democracy. Given that our brains have evolved to respond to immediate and not far-off threats,[394] we will first have to find means by which to proactively renounce the rapid results that an excessively efficient political system may achieve. On an institutional level, this could take place by means of the creation of a third parliamentary chamber in the long term, appointed to discuss environmental challenges and social robustness. It is this solution in particular that philosopher Dominique Bourg proposes in *Pour une 6ème république écologique* (For a 6th environmental republic).[395] The Citizen's Convention for Climate may be the prelude to this.

6.13 DECENTRALISATION – APPROXIMATION

As discussed previously, living organisms are largely self-organised. This is made possible by a modular architecture, whether in structures (organelles and cellular compartments, cells, tissues etc.) or in functions (metabolic modules, signalling modules etc.). This makes it possible to contain local failures (as in the autopilot system of aircraft, which uses different redundant modules). This modularity allows adaptability: a neighbouring cell can take the place of another; a metabolic pathway can compensate for the deficiency of another pathway.[285] This robust self-organisation through local modules invites us to consider the rise of decentralisation in social systems.

We saw with Elinor Ostrom how the robust management of the commons involves self-organised governance at the local level. The Anthropocene invites us to revisit our history, the role of the commons, of the local and of decentralisation. For example, from the 12th to the 18th century, the enclosure movement gradually converted the commons into private properties in England. This destruction of the commons within the country itself and to the benefit of a few rich landowners prefigures the centralised capitalism of the Industrial Revolution. For the philosopher Isabelle Stengers, this appropriation of land, this "mutilation", is even the domestic side of colonisation.[396] Today, such a trajectory cannot be sustained anymore. As Thomas Piketty puts it, "the hyperconcentration of ownership and power does not correspond to the needs of a modern and circular economy".[397]

Besides the economic standpoint, a return to the commons and decentralisation could also involve psychological foundations. Experiments have actually shown that the capacity for cooperation in managing a common good decreases as the size of the group increases. In other words, a common good will seem all the more precious when the size of the population that exploits it is small. The fragmentation associated with decentralisation can therefore be a mobilising force for populations. Note that, as with donut theory, there is also a lower limit, that is, a minimum population size that would have the human means to be able to manage this resource.[398]

Decentralisation thus indicates that local responses can be engaging. Paradoxically, a partial (and thus far from the ideal of scientific objectivity) but more intimate knowledge can generate new capacities to respond locally, to recognise one's duties and debts. In a form of continuity with the previously mentioned solidarists, and to use Donna Haraway's term, situated knowledge can create the conditions for "response-ability".[193] In other words, in the scenario of situated knowledges and practices, many ingredients of suboptimality – heterogeneities, incompleteness, inconsistencies or redundancies – would be present, with robust resource management as an emerging property.

Is this scenario utopian? A look at recent history seems to predict a trajectory towards decentralised cooperation: contemporary societies are becoming increasingly horizontal. This is a passive mechanism, brought about by the increased power of the socio-economic world over politics, and more recently amplified by the digital revolution. Jérémy Rifkin even speaks of the Third Industrial Revolution as a means of describing a world that has transitions from a clear verticality of power to lateral interactions within a network.[399] In its various scenarios for 2035, the CIA also envisages a world of communities where foreign policy and defence will still be entrusted to the state, while healthcare, education and economy will fall under the sole jurisdiction of local regulatory authorities.[364]

Nowadays, decentralised territories have acquired much greater control over their own destinies than ever before. The experience of the "Zone to Defend" (Zone à Défendre, ZAD) in Notre-Dame-des-Landes (near Nantes, in France) is an extreme case: the State is almost absent within a territory that has been "taken" by the Zadists. For example, there are no police in the ZAD, even though it is located in a law-abiding state. Such an autonomy has led the citizens of the territory to find new ways to manage the lands, resources, conflicts and so on resulting in the formulation of strategies by which to replicate this type of experiment elsewhere.[400] This represents a genuine total democracy experiment. Now within a more institutional and less conflictual context, entire territories, such as the Biovallée in Drôme, have decided to form federations of communities and associations, united around the issue of sustainability. An increasing number of towns and cities are developing urban and peri-urban agriculture in order to guarantee their supply of local and often organic produce, and to monitor the quality of their water.[401]

In addition to essential infrastructure and services, this decentralisation also affects trade. For several years, the city centres in Europe have been renovated, in particular with the aim of responding to the problem of shop vacancies. This is in fact a late reaction to the hyper-performance associated with large, outlying commercial centres: these gigantic consumerist machines have generated profits in the short term; however, they have destroyed the urban and peri-urban social fabric in the medium term.[402] The French city of Mulhouse could serve as a model on the basis of the revitalisation of its city centre: having a record number of independent local shops, this city has become very attractive to residents, tourists and businesses. This dynamism is the fruit of the greater heterogeneity of the more unpredictable, perhaps even more incoherent commercial offering, and it results from the criticism of the so-called performance of the major chains.[403] Similarly, the Spanish city of Pontevedra has aroused curiosity and won awards from around the world when its mayor decided to exclude cars from the city and give it back to pedestrians. In addition to changing infrastructure and regulations, this shift to pedestrian rule also involved resisting the implementation of peripheral shopping centres. Thus, this genuine systemic approach also included an indirect support for shops in the city centre. Since 1999, the success of this strategy is undeniable.

Historic circumstances can also drive cities to a form of positive suboptimality. This is the case in Berlin, for example, with its abundance of alternative cultures that have coexisted during the 30 years since the fall of the wall in 1989. As journalist Ann-Kathrin Hipp nicely phrased it in *Der Tagesspiegel*, “the city appealed because it was imperfect and lent itself to approximate lifestyles”.[404] A major challenge for Berlin, during a phase of accelerated gentrification, lies in finding the means by which to preserve this ever so adaptable and welcoming spirit in future years.

Ultimately, decentralisation is a twofold approximation. In the mathematical sense, it is a matter of allowing a deviation from the average, that is, a form of imprecision that respects and values situated knowledges and local contexts. Etymologically, it is bringing people closer, reconciling authorities and citizens. In this context, villages and districts would perhaps represent the appropriate scale on which to respond to the environmental challenges, simply because the visible effects can be readily understood by the directly impacted local community. In return, the solutions provided are manageable, and avoid the often-paralysing feeling of powerlessness. Mayors and citizens are increasingly taking hold of this suboptimal power, even exercising civil disobedience, notably on the matter of pesticides. It is safe to say that with climate and ecological fluctuations becoming more intense, municipalities will be on the front lines in the future. We should therefore anticipate this trajectory now, in particular by training the next generation of mayors to harness this power, and by creating means by which to get citizens more involved.[405]

On a broader scale, national research institutions such as INRAE are investing in *living lab*-type strategies that bring together researchers and local populations. The Fondation de France finances *new patrons*-type projects (“nouveaux commanditaires”) that call for artists to respond to local issues at the request of the residents of the respective territories. By no means merely anecdotal, 500 new projects have been realised within the Fondation de France’s 25-year existence, with the foundation providing average funding of 150,000 euros per project.[406]

More generally speaking, the revolution of the autonomy and decentralisation of territories could once again provide a response to a more profound causative factor: the pressure placed on resources that are nearing depletion. A state can be allowed to control everything centrally, provided that economic growth muffles discontent and dysfunction on a local level. When crisis and scarcity appear, decentralisation naturally emerges. The state opts out and makes use of the remaining resources to focus on its sovereign duties.

However, there is a significant risk that we might passively suffer this evolution rather than actively anticipating it. For example, with the depletion of resources, some social services will be impacted extremely severely. This is particularly applicable within the healthcare sector, which is reliant on oil for several products, for the transport of goods and people and for the operation of equipment. The short channel concept should thus be applied to this sector as a preventive measure, especially in more fragile countries, so as to anticipate the consequences of peak oil. For example, the utilisation of glass that can be re-used and sterilised on site must be resumed, local equipment and materials favoured, and easily mobilisable, non-motorised means of transport identified. Looking beyond logistical aspects, this would involve a profound change in the governance of healthcare systems, with increased accountability for local players, for example, through the creation of a public health centres in each village.[407, 408]

Why should we not, in contrast, recentralise power in order to control expenditure and better cope with international competition? Let us look back to the early days of the Industrial Revolution, when this centralising trend first emerged. At the beginning of the 19th century, when the government of the United Kingdom decided to abolish multiple local regulations pertaining to factories in favour of single centralised regulation, it based its decision on the abundance of available resources and the promise of future growth.[35] The pre-existing autonomy of territories was muffled. The link between the scarcity of resources and decentralisation is again apparent, but now in reverse. Needless to say, this centralisation of power in the United Kingdom in the 19th century enabled and maintained huge laxity with regard to pollution and environmental regulations. Luckily, this digression is currently coming to an end, and Great Britain is actually leading the way, with zero-energy towns and villages.[409]

I can hear another voice: Doesn't digital technology allow us to decentralise services? Wouldn't it allow us to make the most of local knowledge while optimising access to it? In its current design, digital technology is a clumsy form of decentralisation. It is too dependent on non-renewable resources and delocalised expertise to be truly decentralised. Digital technology is dependent on Chinese rare earths, on a limited number of giant data centres and on the stranglehold of Silicon Valley players. One could say that the digital revolution is instead a formidable centralisation power.

But let us be fair, let us consider the digital without its physical and geopolitical base, and consider its services, which are indeed decentralised by construction. In 2010, the Google Flu Trends web service in the United States cross-referenced the occurrence of terms summarising flu symptoms in local conversations, based on geo-localised Big Data on the Internet. Adopting a bottom-up approach, this artificial intelligence detected a flu epidemic 2 weeks prior to the CDC (Centers for Disease Control and Prevention), which requires top-down validation.[410] However, could the increased efficiency made possible by extended decentralisation thanks to digital technology set us on a direct course towards the inadequacy of optimisation? The Google Flu experiment was not pursued in its current form: significant bias was detected in the results, such as the over-forecasting of flu.[411] The delays of the CDC thus proved necessary. We have here a typical case of hyper-optimisation, whereby the negative effects surpass the aforementioned benefits. A limit of decentralisation discretely emerges in its most advanced form: it can be confused with personalisation. It is therefore a matter of decentralising up to a certain threshold, or in other words, of *hindering* personalisation. Again, it is a question of giving primacy to robustness over performance. Which suboptimal pathway could play such a restricting role?

6.14 SAY NO TO HYPER-PERSONALISATION

Suboptimality highlights the conflict between individual comfort and communal robustness. It thus calls into question the current wave of individual hyper-optimisation through the personalisation of services and products, by means of marketing, digital technology or artificial intelligence. Is this hyper-optimisation failing?

The negative effects of hyper-personalisation are in fact very widespread, and they are particularly evident within the digital revolution. For instance, aside from the biases of Google Flu in forecasting epidemics, artificial intelligence that decodes language is generally sexist and racist, simply because this bias exists in the content that human beings publish online.[412] Hyper-personalisation mainly promotes a form of communication without filter. The example of flash crashes on the stock exchange, caused by algorithms that are faster than human brains, must be taken for what they are: the preview of a future world subjected to widespread over-optimisation.

Cambridge Analytica offers another well-documented example, this time within the political sphere. This company collected the personal data of millions of American voters on Facebook, and used this data to carry out *microtargeting*, that is, the transmission of personalised campaign messages to a previously unparalleled granulometry. In fact, it is now well established that a small number of *likes* on social networks is all it takes to determine our main character traits, provided that the right algorithm is in place.[413] When this type of tool is placed in the hands of extreme right-wing reactionaries, it becomes a vector of manipulation. For example, it is possible to send "*dark posts*", that is highly personalised messages that pass for public messages, sent to people deemed likely to switch political camps. In addition to the content of the message, the time at which it is sent is also personalised in order to guarantee that it will be read, and at the optimal time of day. Cambridge Analytica thus specifically targeted undecided voters in the American states with the largest electorate, contributing to the following result: Donald Trump was elected.[414] In spite of these warnings, it appears difficult to battle against a power as strong as individual comfort, made possible by means of ever-increasing levels of personalisation.

One could delve even further into the catastrophe scenarios associated with hyper-personalisation. In the aforementioned examples, the default context envisaged is that of peace. What would happen in a time of war? Would it be possible to mount resistance movements within a hyper-personalised, and therefore hyper-surveilled society? The history of every nation is rich in these redemptive moments of insurgency against the excesses or abuses of power. Suboptimal monitoring has its shortcomings. However, in the long term, this weakness enables resistance, and consequently a form of societal adaptability and transformability – a collective resistance. Finally, an element of suboptimality could feed a reaction to the "control society", built on a far-reaching decentralisation of "disciplinary institutions". The writings of French philosophers Michel Foucault and Gilles Deleuze on these themes are gaining new vitality as digital data management shows its true face.

If suboptimality calls into question the value of such hyper-personalisation, what would a better balance between individual comfort and collective robustness look like? Has personalisation already been proactively thwarted? We have seen how rights, regulations and taxation form obstacles to personalisation. Are there more recent examples where the benefits associated with suboptimality have been implemented, as opposed to the pursuit of personalisation?

One does not have to look far: consider the book that you are holding in your hands and its price. A product of economic liberalism in the 1970s, the free pricing of books in many countries still follows the rules of freedom of trade: the same book will vary in price if purchased in a local bookshop or in a supermarket. This creates a tension within the entire book production chain, on account of stiff competition. In contrast, the Lang Law of 10 August 1981 instituted fixed book pricing in France: a book will be sold at the same price everywhere, resulting in the *de facto* restriction of freedom of trade. Not only does this regulated economic model prevent competition between retailers but it also creates solidarity within the entire book production chain, from author to reader. This is a cultural policy, whereby suboptimality is understood and assumed. In 2018, 14 countries had put in place such legislation.

There are in fact many examples where non-personalisation is asserted. We have seen how agroecology asserts a form of imprecision, as opposed to hyper-personalised precision agriculture. Similarly, prophylaxis, a set of measures implemented with a view to preventing a disease, is suboptimal, in particular because it is generally not personalised. Vaccination is the archetype: on a population-wide scale, it is the total immunisation coverage that counts, and not whether any specific individual is vaccinated. Thus, concerning measles in France, the objective is to reach 90% vaccination coverage at 6 years of age with a view to eliminating the disease.[415] With the promise of customised care and personalised medicine, enabled through the use of digital technology and biotechnology, prophylaxis, and its associated suboptimality, may be called into question in the future. There remains hope that this will not be the case: future pandemics, similar to Covid-19, may remind us of the benefits of vaccination.

After decades of free and fair competition paradoxically producing legislative complexity, our contemporary societies are beginning to reflect on the implementation of simple universal rules, and on their systemic effects. Rules are the basis of life in society, and they are not personalised: they are imperfect from an individual point of view. In fact, if there were as many rules as there are citizens, these rules would not be known or accepted, and therefore they would not be applied. This is a question of double robustness: (i) the social rules allow the survival of the group in the long term and (ii) the simple formalisation of these rules allows their understanding and application. This simple and suboptimal principle is constitutive of human societies.

It is as though our new relationship with the planet in the Anthropocene has obliged us to stand back and impose more universal and clearer rules. As we have seen with agroecology and decentralisation, it is not a matter of homogenisation, but rather universality, providing a fertile setting for an underlying heterogeneity. This is illustrated in secularism, which does not prohibit religion, but rather permits the practice of all religions. As philosopher Francis Wolff expressed in his *Plea for the universal (Plaidoyer pour l'universel)*, this shared common framework is not a censor, but rather a mobilising element.[416] This form of engaging and liberating simplicity may come to be applied more widely. It is now time to address the matter of sobriety.

6.15 SOBRIETY

One of the keys to our coexistence with Earth is sobriety. This is actually one of the main conclusions of the Meadows report. The only scenario that would avoid painful turbulences involves a form of total sobriety: proactive stabilisation of demography, reduction of material production, use of technologies limiting environmental impacts.[37] Within a society that will soon be exposed to the exhaustion of resources, we will be required to reduce our consumption and therefore our needs. The objective is undeniable. Sobriety rather raises a question of strategy.

We have seen how digital technology or nanotechnologies, intended to lead us to frugality, instead do just the opposite. At this point, I will not repeat myself regarding the extraordinary, and growing, costs in terms of the energy and resources associated with digital technology infrastructures. Neither shall I return to a discussion of the Jevons paradox, which reflects the curse of efficiency. To optimise products in order to construct a sober society would be naive at best.

Instead, it is urgent that performance be reduced, and therefore that "disinnovation" occurs within certain sectors. As Philippe Bihouix wrote in the bountiful book *The Age of the Low Tech*, the field of opportunities is immense.[123] For example, we could standardise high-consumption items in order to combat their planned obsolescence and promote their local re-use. In fact, we already know what to do for common products, such as returnable bottles. We could also regulate competition, obliging the sharing of common infrastructure (e.g. between telecommunication providers). Going further, it would simply be a matter of ceasing the manufacture of certain products, either because they are not absolutely necessary, as they contain pollutants, or because the resources required in manufacturing them should instead be invested into more useful products. Is there really a need for radio-frequency identification (RFID) chips in clothing? Showerheads with multicoloured light emitting diodes (LEDs)? Can the intercontinental shipping of mineral water or yoghurt still be justified in this day and age? Was it really necessary to use 2,000 tonnes of plastic to celebrate Halloween in the United Kingdom in 2019?[417]

Sobriety, through robust strategies, also presents a huge opportunity for researchers. Being the driver of economic growth, innovation and therefore applied research, often remains fixated on efficiency. To shift the compass of research towards robustness is a profound revolution, which opens up a great many new questions. For instance, one major challenge lies in producing goods that are reusable and integrated into the Earth's cycles. Taking into account the aforementioned limitations, the circular bio-economy is without doubt a significant lever. But it is not enough. Social robustness must also be taken into account.

Let us take a very concrete example. In the performance society that has become aware of environmental issues, the electric car is often held up as a standard of responsible purchasing. In addition to the negative externalities associated with battery management and the many rebound effects (such as the rise of the electric sport utility vehicle [SUV]), the complexity of these new technologies makes citizens captive and dependent consumers. The socio-ecological outcome of the electric car seems ambivalent, at best. So, what form would the car take in a society that aims for sobriety and robustness? First of all, it would be light, shared and repairable. Whether the engine is thermal or electric would be secondary. In fact, maintaining a certain technical diversity (petrol, diesel, gas, hydrogen, electric etc.) would certainly be an interesting target. Such redundancy could, for instance, help public services to cope with the future fluctuations of energy supply.

A sober and suboptimal world is therefore not technophobic on principle. On the contrary, it is in favour of diversifying technical developments and learning. High tech is lazy, because it is built only on performance increments. Low tech is much more ambitious in its objectives, not only because it can call on advanced technologies as in high tech but above all, because it implies an integration of its activities and products into the Earth's cycles. Sobriety, low tech and suboptimality come together: it is rather the sense of technology that is put back on track for the robustness of society.

More generally, sobriety implies the monitoring of our individual comfort, occasionally accepting a service that is less personalised or less fast, in favour of robustness or collective benefit. This could not be more in keeping with suboptimality. Far from being a sacrifice, we have seen how hyper-optimisation created needs, frustrations and ultimately result in a stalemate. What is actually required is that we dare to develop the emerging properties of less personalised or slower services.

Sobriety is therefore not a simple reduction of consumption and waste. Rather, sobriety is a set of conditions that place us in a situation where consumption and waste are no longer problems. In other words, suboptimality allows us to move from a punitive ecology to an engaging robustness.

In particular, sobriety under the suboptimal flag emphasises repair and reuse. This does not only refer to material sobriety. The most interesting emerging property of repair and reuse is the necessary intermediation between citizens. Unlike the acquisition or recycling of goods, which often involves specialised and distant infrastructures, repair and reuse can be managed by citizens themselves in a local and decentralised way. This is happening with the development of second-hand markets, both real and virtual. Similarly, networks of bicycle repair shops not only provide functional bicycles but more importantly, they promote education in knowledge sharing and the commons. If robustness is our new target, it is a safe bet that every town and village will have their repair hubs, just like a post office or a town hall today. As philosopher Victor Petit puts it, it is a matter of moving from the mirage of *smart cities* to the much more robust and engaging perspective of *smart citizens*.[418] This will require efforts to learn to cooperate and build together.

Cooperative housing could be a very advanced form of this new path. In this case, the inhabitants, or rather the co-operators, design, create and manage their dwelling as a common good. In addition to private spaces, there are shared spaces, such as a guest room, a floor for children's play or a vegetable garden. Here again, in addition to saving space by sharing, learning to live together is a factor of robustness. The experience of the Covid-19 pandemic demonstrated the robustness of these habitats, both economically and psychologically. And the emerging properties of cooperative housing spread beyond the walls of the building. When residents participate in the construction or management of their homes, they often also take care of their neighbourhood. Completion and co-construction educate and legitimise a systemic vision of housing. With a unique experience of cooperation, with its difficulties, its slowness, its conflicts, these inhabitants are certainly the best equipped to share their experience and move forward in the century of robustness. Once again, this is not a marginal utopia, but the real world of today and in the making. With 15% of the housing stock in Scandinavia, cooperative housing is a model that is already proving its worth and is expanding. Choosing sobriety through collective action is claiming the virtues and emerging properties of the infinity of interactions.

6.16 A SOCIETY OF ABUNDANT INTERACTIONS

Suboptimality, as a response to the overexploitation of resources in the Capitalocene, could echo the theory of degrowth. Again, this concept suffers from a multitude of definitions. Here is the one I discuss: the objective of degrowth is to bring about a sober society by reducing the use of materials and energy. Degrowth first and foremost concerns the most greedy economies (i.e. the Organisation for Economic Co-operation and Development [OECD] countries), and would target the economic sectors that are, environmentally speaking, open to criticism and offer little social benefit (single-use plastics, SUVs, marketing, etc.).[127, 419] Degrowth means "less but better".

Besides the link to sobriety, in what respect could degrowth echo suboptimality? Reading (too) quickly, degrowth might resemble a recession imposed from above, producing results on an ecological level, but at a questionable socio-economic cost. Instead, degrowth corresponds to a more systemic paradigm shift. This would involve constructing an economy that no longer requires growth in order to exist. Degrowth is only a transition phase towards *post-growth.*

Within a degrowth scenario, the focus would no longer be on performance, but rather on the indicators of development and human well-being, while remaining respectful of the planet. Degrowth is mainly a European variation of a much older and wider movement on a global scale. Examples include Buen Vivir ("good living") in Latin America, Swaraj (स्वराज, "emancipation") in India or the more recent Tang Ping (躺平, literally "lying flat") in China. This deserves a side note.

Buen Vivir is an ancestral philosophy, inherited from pre-Columbian Indigenous communities. This school of thought preaches harmony between humans and non-humans, placing the relationship with Mother Earth at its heart. When Rafael Correa's Ecuador incorporated Pachamama into the constitution in 2008, he supported the Buen Vivir movement by understanding its systemic effects. Indeed, the concept itself is very robust, having survived the conquistador colonisation and subsequent cultural collapse.[420]

In the case of the Hindu Swaraj, it is rather a reaction to an invader, the Muslims in the 17th century, and more recently, the British colonists. Mahatma Gandhi is often presented as the figurehead of this movement. The Swaraj fully assumes its political role by emphasising decentralisation, and thus self-governance.[421] Again, the emergent properties of such a philosophy create robustness. So much so that it defeated the British army during the Indian independence movement.

Finally, Tang Ping appeared on social networks and is developing among young Chinese who refuse to follow the injunction to work hard (the 996 model: work from 9:00 am to 9:00 pm, 6 days a week). Tang Ping is a form of resistance to a society that no longer inspires dreams and is in relative decline and disenchanted.[422] It resonates with young people around the world. Some people even speak of the global movement as a "collapse generation". It could actually be related to the romantic Germanic concept of *Weltschmerz*, a form of depression in the face of a suffering world. And a personal inadequacy to inhabit it.

One might think that Tang Ping is less engaging than Swaraj or Buen Vivir. Yet stopping is the first operational step towards a change in society. As researchers Emmanuel Bonnet, Diego Landivar and Alexandre Monnin write in an opinion piece, "closure is the most optimistic political horizon of the new climate regime".[383] Or how to go beyond the limits of the mind-numbing discourse of collapsologists: questioning our trajectory, or even stopping it, is an essential step before building a new world.

Buen Vivir, Swaraj and Tang Ping invite us to reassess the measure of progress. In the 1990s, the Human Development Index (HDI) was created as a means by which to measure social progress, addressing the limitations of GDP, which measure only economic progress.[423] For instance, the ten countries with the best HDI include the Scandinavian countries, Switzerland, Australia and Singapore. However, these countries also have a very large environmental footprint per citizen, which would in fact condemn all human development in the long term. Exemplifying the ongoing shift from the social contract to the natural contract, the new Sustainable Development Index (SDI) now aims to address this mistake by dividing the HDI by an environmental impact index. The list of the ten countries with the highest SDI is thus very different, with Cuba, Costa Rica and Sri Lanka on the podium. None of the countries with a high HDI is represented. Norway, for example, lags behind in 157th place.[424] For information, the United States are in 159th place, the United Kingdom in 131st place, China in 100th place, France in 95th place, India in 55th place.

The new choice of indicators is not just for show. It reflects a deeper change in the very notion of what is activity, as we have seen previously with the new status of work. Given the necessary conversion of old industries, the decontamination of sites and the redevelopment of cities and territories, the working needs and opportunities are growing. Some even speaks of "negative commons" to describe the many remains of the Anthropocene, which are becoming useless and that we will have to manage in the future.[425] According to scenarios devised by several economists, this rising degrowth economy could emerge and be viable. It would require the conversion of the jobs of the old growth economy into more virtuous activities, the implementation of a form of universal income and the reduction of the number of weekly working hours.[127, 426]

As is the case with counter-productivity, the idea of degrowth is based on the observation that there exists a threshold beyond which the indicators of social well-being decrease as GDP increases. Thanks to the investment in their healthcare system, the citizens of Costa Rica now have the same life expectancy as those of the

United States, in spite of having a GDP per citizen that is five times lower.[427] Europe has a GDP that is 40% lower than that of the United States; however, its social indicators are more favourable in all sectors. Thus, there is another Kuznets curve: economic development, environmental impact and impact on social health are initially synergistic, then they reach a plateau, and finally they become antagonistic. The cost of advanced economic development has negative effects on the environment and on social health, which ultimately outweigh the gains offered by wealth alone.[127]

Growth is often presented as the fuel of a society of economic abundance, thus condemning the degrowth society to the image of scarcity. This reading confuses the quality and the origin of abundance. Within the modern growth economy, material abundance results from the privatisation of certain territories and services, without a genuine connection to the physical world. Scarcity is created in this manner to generate need and frustration, and therefore to establish a market. Although seldom acknowledged, this is actually the main goal of advertising. The transition from a subsistence economy to an export economy during colonialisation is a clear example of this swing towards an economy of artificial scarcity.[204] Planned obsolescence is another illustration of this. Economic growth showcases abundance, but fundamentally it constructs scarcity, whether real or artificial. By emphasising common goods and by considering planetary boundaries, degrowth bases its economy on more real foundations. It puts an end to the artificial needs created by virtual shortages. It is thus built on the authentic needs of society. These are channelled by constraints that are multiple, realistic and connected to the planet. Degrowth therefore clearly distinguishes sobriety and scarcity. The ideology of growth is based on scarcity, organised or genuine. Although founded on the objective of sobriety, degrowth is in itself founded on abundance: the abundant potential of interactions. Finally, with this definition of degrowth, it is about creating value. Not by exhausting the Earth's resources in the phantasmatic framework of material infinity, but by taking advantage of emerging properties in the framework of the infinity of interactions.

Growth and degrowth are also opposed in their relationship to time. The growth-focused model favours long weekly working hours and creates a variety of needs: time is in continual shortage. It is the other way around with the degrowth scenario, which generates an abundance of time to facilitate interpersonal interaction. As we have seen with the Social Solidarity Economy, this opens up a window of adaptability with regard to the real problems faced by society and the planet. Incidentally, it can be noted that the 4-day week is increasingly being put forward as one of the solutions to the climate crisis. For instance, a report by the Autonomy think-tank flips the working time paradigm on its head: "Rather than discussing how to maximise economic performance, the climate crisis forces us [. . .] to change the question: given the current carbon intensity levels within our economies and our productivity levels, to what extent can we still allow ourselves to work?"[428]

Once again, the range of possibilities can open up, not by means of the desperate exploitation of resources but instead by taking advantage of the interactions between Earthlings in a society that coexists in and alongside ecosystems. It actually involves a broader change in mentality. In the past, we used matter to save time (e.g. jet fuel for flying). In the future, in the sober circular bio-economy, we will rather use time to preserve matter. In the so-called conventional agriculture, we were exploiting ecosystems in order to increase the productivity of cultures. With the rise of agroecology, our agricultural production will feed ecosystems. We were living in a growth society based on artificial shortages. We will build a sober society based on abundance and complexity of interactions. Finally, by setting the rails of progress on robustness, not on performance, suboptimality leads us to question our socio-economic models and our organisations in depth.

6.17 LOOSELY COUPLED SYSTEMS

The infinity of interactions and their systemic effects on robustness can remain theoretical and difficult to grasp in day-to-day life. So let us now head to the office or to school, and explore the suboptimality of organisations.

Multiple interactions, whether hierarchical or otherwise, between authorities and employees, between employee and work tool, or between professor and student preserve unity within organisations. These interactions, or couplings, have varying degrees of flexibility. For example, a "rigid" coupling between an employee and their hierarchical superior usually is direct, significant, long-lasting and therefore predictable. For example, the superior gives an order, and the employee is expected to respond efficiently. In contrast, a "loose" coupling is indirect, sometimes irrelevant, discontinuous and therefore largely unpredictable. One might think of coffee corner discussion, where any notion of efficiency, or even efficacy, is absent.

This distinction has been the subject of many formal studies, most notably through the analysis of *loosely coupled systems*. There are several definitions associated with this concept, which can result in confusion.[429] Here I define loosely coupled systems as complex systems, whereby the elements retain a degree of autonomy, and therefore unpredictability. This definition is not far away from that offered by organisation theorist Karl Weick, who introduced the concept in a seminal article in 1976.[430]

Weick built his study on the analysis of the educational system in the United States in the 1970s. He observed that this system presents a particularly high level of flexibility: an excess of free time, resource redundancy, a lack of global coordination, absence of regulations, a high level of independence, etc. Although some might raise the alarm and be tempted to "rigidify" such a system, Weick instead highlights the virtues of such flexibility.

First and foremost, loose couplings enable a more acute perception of the environment. Let us consider an analogy. Imagine a storm warning system based on the observation of trees. In order to measure the speed of the wind, all you have to do is watch the flexible branches of a tree moving. If we could only see the very rigid trunk, our information would be very limited. A degree of flexibility in couplings is necessary in order to perceive fluctuations.

Furthermore, loose couplings promote robustness. To continue with the plant analogy, this is a variant of the fable of the Oak and the Reed: a loose system will be able to adapt better to environmental fluctuations. In the case of the educational institutions studied by Weick, the system persists over time, in spite of the many challenges. In the most extreme cases, if part of the system were to collapse, the system as a whole will be less affected if the couplings are loose than if they are rigid. The high level of autonomy prevents the global contamination and destruction of the system.

Finally, loose couplings limit overall coordination, in particular through delays in the interactions and through redundancy. This enables adaptability. In particular, the lack of global coordination facilitates the establishment of local initiatives. Conversely, this type of organisation is less susceptible to synchronous and systemic change, given the high level of intrinsic autonomy made possible by loose coupling. In the end, loose couplings uncouple local and global scales. This creates robustness: the global system becomes less sensitive to local disturbances.

Some elements of suboptimality have already been set out. Yet the question as to the desirability of suboptimality also arises: loosely coupled systems can also have negative effects in their most extreme forms, such as excessive sensitivity to fluctuations (e.g. fads) or resistance to global change. Nonetheless, the more counter-intuitive benefits of these loose couplings remind us of the adaptability and robustness associated with suboptimality in living systems.

Let us take this a little further: Is it possible to find initiatives that proactively use loose couplings within contemporary organisations? The success of a company often depends on the ability of entrepreneurs to identify their true goals and to ask the right questions, even if this means departing from the dogmas of rational management. This includes the idea that local optima are not necessarily conducive to global robustness, as in living beings.[431] The work by Frederic Laloux, *Reinventing Organizations: A Guide to Creating Organizations Inspired by the Next Stage of Human Consciousness*, illustrates how some organisations are in the process of voluntarily distancing themselves from a method of management that is too efficient and too rigid.[432]

The example of care facilities in the Netherlands provides us with a clear understanding of the transition from an engineering-style optimalist model to a more suboptimal model. In the majority of modern care facilities, nurses are grouped together and organised via call centres, as is the case in assembly lines. They provide treatments, sometimes referred to as "products", the duration of which is regulated and scheduled. In some cases, nurses even go so far as to scan a bar code, which patients are then required to stick to their front doors in order to optimise flow management. The next increment in this drive for efficiency would probably be the invention of robot nurses or the performance of remote treatments without patient contact. However, before we even reach this level, optimisation reaches its logical dead end as a result of the growing discontent of patients and nurses, and as a result of poor medical results. A typical case of counter-productivity, where negative feedback prevents the achievement of the desired results.

The Buurtzorg Nederland organisation was created in 2006 by Jos de Blok in direct response to the limitations of this over-optimised model. The organisation of care treatments supports patients by listening to them, by putting them in touch with their neighbours, and by taking the time to really understand their environment, their needs and desires. The medical results are excellent, since the detection of illnesses is much better anticipated. Rather than costing the Dutch health system more money, this organisation renders patients more autonomous and even generates savings for the state.[432] The medical and financial success of this organisation has seen it spread throughout Europe, but also to the United States and in Japan. These organisations provide an operational response to the impending increase in dependency, associated with the ageing of more "developed" countries.

On an internal level, Buurtzorg Nederland resembles a self-organised structure, without hierarchical levels. It functions as a complex system where local interactions dominate and give rise to global properties, without any genuine need for top-down control. This is remarkable given that this organisation now numbers 10,000 members. Governance is distributed: any member can request the opinion of external experts and the opinions of those persons affected by decisions. This is still a complex system approach, since decisions are not made in a hierarchical manner, nor based on a global consensus: decisions are reached as a result of local interactions. This type of management also enables a high level of adaptability. Each member or group of members can actually innovate without formal planning, and without hierarchical decision-making.

The Buurtzorg revolution can recall the early days of self-management. However, it can be noted that this embryonic form of governance misunderstood the difference between local opinion (as in a complex system, in self-governance) and global consensus (which does not always respect heterogeneity of opinion, in self-management). In other words, the consensus enabled by self-management can remain locked in the dogma of performance. For example, the drive towards consensus may push everyone to find a solution as quickly as possible. On the contrary, the *common sense* (using the definition by Isabelle Stengers[396]), fuelled by self-governance, is built on heterogeneities and antagonisms. Problematising first makes it possible to be more robust over time. This relates again to the nexus of robustness and time: not only does robustness make it possible to envisage the long term but it also needs enough time to develop.

The fact remains that, intriguingly, the generalization of self-management came about at the very moment when the environmental issue was emerging as a topic of public debate. As with digital technology, this simultaneity could be more than a coincidence[148] in becoming aware of the finite nature of the world, of the depletion of resources (material and human); the human being has no other choice than to consider the final window to infinity, interpersonal interaction. In this hypothesis, it would not be surprising to see other cooperative forms of organisation emerge. Suboptimality could provide a contribution in this regard.

However, this picture needs to be nuanced. It is crucial to point out some of the limitations of the horizontal self-governance model, even in the case of Buurtzorg. In particular, the commitment sparked by autonomy can also create new negative collateral effects. The importance attributed to the rules of the group could result in new forms of control that would be difficult to justify. As in *Brave New World*: "And that, said the Director sententiously . . . is the secret of happiness and virtue, loving that which one is obligated to love".[433] The risk of voluntary self-enslavement should not be overlooked. Individual freedom requires a certain amount of ignorance, even indifference, with regard to each another's activities and thoughts. Some degree of suboptimality could serve as a safeguard: for the robustness of organisations, control must not be total. It must leave room for resistance. Finally, suboptimality does not sit well with radicality, which often remains locked in performance and is thus too fragile to claim any kind of robustness.

Because of its ambivalence, this type of management has been misappropriated by certain economic players. This is actually what occurred historically with the criticism of Taylorism, resulting in the promotion of autonomy and initiative to the detriment of the feeling of security experienced by workers.[434] One could say that horizontal self-governance or self-management make sense only if the goal is the robustness of society, and not performance. When horizontal governance is promoted to support performance, it is actually worse than vertical, hierarchical governance, because each member of the organization is now responsible for the success of the entire structure. To be crystal-clear, as shown by historian Johann Chapoutot in *Libres d'obérir*,[435] the Third Reich army was already using self-management principles, with an obsession for efficiency. Self-governance only makes sense with robustness in mind. For instance, a large degree of redundancy would help the members of the organization to feel safe, and would further promote interpersonal interactions. Even if management reinvents itself, it will only be able to claim suboptimality in the future if long-term robustness outweighs short-term performance.

6.18 AN INTRINSICALLY SUBOPTIMAL CULTURE

While certain means of operation within organisations, political systems or communities are in the process of evolving into robust, suboptimal forms, other sectors have never ceased to be suboptimal. This is the case with culture. According to historian Yuval Harari, culture is a system that attempts to preserve its own survival, as would be the case with a living organism: "Cultures are mental parasites that appear by accident and then take advantage of all those that they have contaminated".[343] Let us take a look at this cultural virus.

In a documentary, Georges Bataille goes to the heart of the subject: "One cannot be moved without confusion".[436] An essential function of art is to make us feel alive. Such sensitivity can only exist through an induced emotional fluctuation, which puts us in a position of weakness and confusion. Art exposes, before being exhibited. This also applies to our relationship with Earth. Sociologist and philosopher Hartmut Rosa even makes it a necessity to overcome the shortcomings of the Anthropocene: by conquering the world, we have turned it into something available, diminished and ultimately disenchanted. Instead, art invites us "to make the world uncontrollable".[437] This form of humility could be a crucial vector of reconnection to the Earth.

It will come as no surprise that many markers of suboptimality can be found within the artistic sphere. Heterogeneity (or rather diversity), slowness (or rather long-term work, meditation, inspiration), inconsistencies (or rather provocations, debate), and (claimed) inefficiencies are all markers of the arts. In the end, artists do not care about optimisation. The artistic struggle against optimisation not only demonstrates resistance to the dominating world of control and performance but also highlights all the virtues of adaptability in a suboptimal world.

For instance, referring to the design of social housing, urban planner Patrick Bouchain declares "to make an architecture that has only one destination is a failure".[438] If this theory could raise doubt in some people's minds, the Covid-19 crisis demonstrated the limits of the optimised city. The principles of functional urban planning are limiting the diversity of uses. In pandemic time, we had to turn car lanes into bike lanes, pavements into queuing spaces, squares into restaurant terraces.[439]

In other words, architecture and urban planning in a changing world will mainly emphasise the values of adaptability.

Artistic creation does not consider mistakes as failures, but rather as means. It actually echoes our own trial-and-error learning: our brains memorise mistakes so that they do not happen again. This is even the basis of education for young children. This implies that the desire to avoid mistakes condemns us to ignorance. Some cultures have understood the virtues of errors and welcome them more favourably. For instance, "Card File" (1962) by American artist Robert Morris, takes the form of a drawer with 48 cardboard cards. It exhibits and asserts an inventory of errors, accidents and losses as essential steps in creation. It is a manifesto against a certain academicism that would prefer to hide the creative process in order to put forward the finish, and probably also the finitude, of the work.

There is also an element of suboptimality in our interaction with the artistic work. For instance, Pol Bury's kinetic reliefs with their random movements are not a hymn to machines; on the contrary, they are a wonderful way for the viewer to experience waiting, long-time, meditation. A sketch is not a sterile incompletion. On the contrary, it leaves enough space for interaction and imagination. In fact, the whole process of making the work of art calls for suboptimality. For example, the social sculptures of French artist Thierry Boutonnier invite collectives to create a city that is as close to life as possible: "Taking root" thanks to the patronage of plantations, distilling a locally produced rose water, etc. From these unexpected interactions, a new feeling of situated conviviality emerges, which gives meaning to the society of inhabitants.

Creativity, which is open by nature, does not sit well with an optimisation that tends to lock it up. It can even surprise the artists themselves. For instance, in a master class, actor Denis Podalydès talks about the moments of discouragement in rehearsals, and how they can paradoxically allow creative energy to emerge.[440] Once again, creativity is not the product of optimisation. Rather, it comes from confusion and a reaction to confusion. In other words, in the words of the National Council of the Resistance, "to resist is to create".[441]

From the interpretation of a piece of classical music, which can never be replicated, to the unbridled improvisation of free jazz, randomness, errors and heterogeneity are all examples of the ruggedness that feeds originality and creates opportunities by which to make the music evolve. Jazz, in particular, provides a framework for the unplanned in improvisation. One might retort that to become a jazz musician demands a certain mastery of one's instrument and of musical codes. Isn't jazz actually the ultimate form of performance? It is all a matter of perspective. The optimalist practice of jazz is in fact possible. Like a young apprentice incorporating the best licks of the most famous jazz musicians into their playing, artistic mastery enables the avoidance of the unforeseen and can thus give jazz a perfectly polished appearance. A return to the military-inspired marching bands! In contrast, the suboptimal practice of a jazz musician brings into play a level of artistic mastery that can welcome the unexpected, thus opening up the infinite nature of music. The practice of jazz also illustrates the difficult balance to be achieved between horizontalisation and hierarchy. As in any form of improvisation, it does not suffice to rely solely on chance for the music to make sense. A key factor in jazz improvisation is intention.

Thus, if the musician were to impose upon themself a technical constraint (e.g. to play the third and seventh at the start of each bar) or an artistic constraint (e.g. to play as if it were very cold, or in the style of Billie Holiday), the errors, delays and heterogeneity that would appear within this musical game will be integrated into this intention. As a nod to viability in economy, there are several possible paths of improvisation, however, this falls within a scope of possibility defined by the framework of constraints.

In spite of this rather optimistic picture of a suboptimal culture, which is perhaps even proud to be just that, we must be cautious in order to ensure its future development: let us not underestimate the influence of the dogma of economic performance on culture. For instance, simplification in the Anthropocene has expanded from food, to the majority of our cultural needs. Major American brands now exist on a planetary scale. In 2019, in France, for example, 55% of theatrically released films were American and Netflix accounted for one quarter of Internet traffic.[442] A small selection of economic interests dominates perspectives and global political and cultural decision-making. The so-called dematerialisation does not lead to decentralisation, quite the contrary.

If there exists a sector in which adaptability and transformability should be preserved, it is indeed culture. Given the structural role of suboptimality within this domain, it is likely that the cultural scene will resist the dogma of optimisation more effectively than other sectors. Oenology incidentally offers a prime example of this resistance. With wines ranging from "délié" (subtle) to "abricoté" (with an apricot taste), this is a sector that enjoys a generous gustatory and semantic diversity. Accordingly, the efforts of wine taster Robert Parker to qualify wines according to standardised, homogeneous and purely quantifiable criteria are opposed by the vast majority of the profession, with some even using the term "parkerisation" to describe a narrow-minded approach to oenology.[443]

6.19 A FLEXIBLE LEXICON

To reflect on an example of the risk of cultural simplification and its response, let us consider language: the most widely spoken language in the world is not Chinese, nor Arabic, English, Spanish or French. It is Globish ("global English"), an inventory of just 1,500 words, in spite of the fact that the English dictionary lists 200,000. In response, the return of local languages and the development of subtle local dialects[444] could serve as a means of cultural rebellion against this oversimplification, and perhaps another sign of the decentralised and suboptimal world to come. This example also provides us with the opportunity to focus briefly on language, which boasts a particularly rich and apparent suboptimality. Among the various cultural forms, language merits a special place as it very broadly defines our identity and ontology. As Japanese novelist and translator Haruki Murakami puts it, "learning another language is like becoming another person".

Language is typically suboptimal. It is built on words added to the dictionary by chance, as it evolves. It features an astonishing number of redundancies. It is heterogeneous in its written form and its pronunciation. Its grammar is rich in exceptions and incoherence. Conversely, language is very robust. Its comprehension is relatively unaffected by errors of spelling or syntax. It is adaptable on account of the fact that it is in constant evolution, and it has successfully passed from generation to generation over a long period of time, even prior to the invention of writing.

Aside from its suboptimal ontology, language has multiple indirect effects on our civilisation and its potential robustness. Chit-chat exists because language is so heterogeneous, redundant and variable. The ambiguities of human language fuel misunderstandings, which in turn stimulate chit-chat. Within an optimalist society, this type of exchange would be considered a waste of time. Incidentally it is largely prevented in Taylorism. In contrast, from a suboptimal perspective, this apparent waste of time becomes an opportunity to construct long-term collaboration, and therefore adaptability. Yuval Harari actually proposes that this human ability would have contributed to the domination of *Homo sapiens* on the planet: the suboptimality of a language enabling gossip enables trust, and therefore trade.[343] This could be another accident of the Anthropocene: within the dogma of performance at any price, not only would we progress contrary to our ontology but also contrary to our history. Therefore, suboptimality would respect not only our intrinsic biology but also our forms of expression.

We also, and perhaps mainly, communicate through non-verbal means. This form of communication is also suboptimal by construction: movements and body expressions are heterogeneous, fluctuating, incoherent, etc. But they are understandable by all, sometimes subliminally. Like the verbal, the non-verbal does not require robotic perfection. Rather, it is this relative imperfection that allows non-expert understanding. This typically suboptimal property is the basis for the robustness of sign language for the hearing impaired.

Let us take this a little further. If, in the Anthropocene, nature and culture are entangled, then we must consider non-verbal language in a broader perspective. This is what Eduardo Kohn proposes in his book *How Forests Think: Toward an Anthropology Beyond the Human*.[445] As in humans, approximate gestures allow for extended communication in animals. But this time, a new barrier is crossed: the suboptimality of non-verbal communication allows communication between different species. For example, when a monkey puckers up its face and shows its teeth, it is sending a clear sign of hostility to its companions, but also to humans who might bother it. It is not the precision of this communication that makes it effective; instead, its imprecision makes it universal.

Similarly, the defence strategies of spiders, which "play dead", show that these semiotic communication strategies have been selected during evolution for a long time. Examples of co-evolution between plants and animals also involve suboptimal forms of communication. With inter-specific non-verbal interactions, we are perhaps taking a new step along the path of coexistence with non-humans. Language, in this broader semiotic dimension, thus enables a form of communication that literally extends beyond humanity.[108]

It is perhaps with communication that we can most easily link biological and social suboptimality. Everything in biology brings about suboptimal communication, whether this is the interpretation of hormonal signals in order to read the environmental light conditions, or the implementation of a communication network by means of root-fungus symbiosis in trees.[446] However, it should be noted that the verbal allows for very elaborate speculation. Its application in writing, on the other hand, authorises a form of memory that goes beyond the oral tradition. The nature-culture dualism thus remains partially operative with two cultural specificities, memory and speculation, which open up the temporal window more widely for humans. In the end, with this *Anthropology Beyond the Human*,[445] our understanding of living together has been extended to non-humans. This may also be an important step in our reconnection with the world, in part thanks to shared suboptimality.

6.20 ERRATIC DIGITAL TECHNOLOGY

While culture and language offer examples of suboptimal forms of expression, they are ancient. It is therefore difficult to imagine how such suboptimality could come about. Therefore, let us explore a more recent language: that of computers and digital technology.

Although it may at first be surprising, this sector might bring relevant comparisons to understanding the emergence of suboptimality. It is not that computers and digital technology are any more suboptimal. Rather it is that this sector is more recent and its history can be easily traced. Furthermore, the associated complexity is better formalised here than in any other domain. It is therefore easier to identify suboptimal behaviour.

The example of self-learning by means of artificial intelligence is illuminating. For example, the computer program AlphaGo Zero has been unbeatable when playing the game Go since 2016, in spite of the fact that no human has ever instructed this machine. If success is accurately measured by performance, the path by which to get there, self-learning, is much more erratic, adopting a trial-and-error approach and thereby containing a degree of suboptimality. The optimalist counterexample to AlphaGo Zero might be Deep Blue. Although it too has acquired the upper hand in chess matches against humans, the expert system Deep Blue possesses intelligence that was programmed by human beings. In other words, Deep Blue has exhaustive deterministic bases where AlphaGo Zero is built on self-learning, and therefore on chance. Given the virtues of suboptimality, AlphaGo Zero is, needless to say, considerably more adaptable than Deep Blue: it has also become unbeatable at chess and at shōgi, whether opposing a human being or any other computer program. Yet when discussing AlphaGo Zero, should we still use the term "program"?

Other examples that are more closely related to our everyday lives show how information technology contains a degree of suboptimality in order to function. For example, the multiple errors and redundancies within computer systems are all sources of global robustness with regard to digital exchanges on the planet. What exactly does this mean? The TCP/IP protocol suite ensures data exchange (e.g. through emails) by fragmenting the information concerned into several packets. These packets are then circulated using the network of multiple machines distributed across the planet. The path followed by these packets is full of pitfalls, and numerous errors are triggered: the loss, corruption or duplication of packets. Incidentally, the path itself is never the same, and can be more or less long. The reconstruction of the information by reassembling the packets also takes time, which can cause delays. The adopted strategy is therefore not that of the optimisation of the trajectory or security of each packets, but rather the robustness of communication as a whole.[447] Internet operations therefore contain a degree of suboptimality.

Blockchain offers another example of robustness built on a form of suboptimality. Digital data must be both easily accessible and highly secure. Blockchain meets this seemingly paradoxical injunction through decentralisation. Data is stored in many distributed accounts, in multiple copies. Updating is done by the network itself, not by a centralised deterministic system. In fact, the structure of the blockchain resembles the many nodes of biological networks, with the associated robustness, and a form of self-organisation through decentralisation. The immense size of today's digital networks is no stranger to this revolution. Indeed, with a limited number of nodes, the system would not be able to achieve sufficient reliability due to a lack of redundancy. Trust emerges from redundancy and heterogeneity, which is itself made possible by the involvement of a large number of actors. Finally, whether it is the Internet or the blockchain, when the value of robustness prevails over the value of performance, it seems that we are converging towards the same recipes as that of living beings.

6.21 APOLOGIA FOR SLOWNESS

To finish on a lighter note, let us now attempt to detect signs of suboptimality emerging in our everyday lives. Whether this lies in the blunders of many cartoon characters such as Franquin's *Gaston Lagaffe* or in the likeable plodding around of *The Big Lebowski*, suboptimality certainly has its fair share of antiheroic incarnations. It is also fully expressed in modern literature. For instance, science fiction novels often deal with the excesses of over-optimisation. Whether in the exhaustive surveillance of *Big Brother* (1984) or in the genetic control of workers (*Brave New World* or *Welcome to Gattaca*), optimisation leads to the dystopia of a homogeneous, risk-free and tasteless world. Fantasy novels also open up to suboptimality, but this time by inventing a universe that is often low-tech, quasi-medieval, more egalitarian between humans and non-humans, and often taking the form of a quest full of random detours. Professor of modern literature Raphaëlle Reynaud even suggests that this suboptimality in fantasy "appears as an alternative, proposing a slowing down of the frenetic race for progress and a detachment from chronology in order to refocus more on space, in synchronicity".[448] Like an ecopoetic response to our diachronic perception of a degraded environment in the real world, to better feel it.

Slowness has also become a political value, notably through multiple forms of resistance to the dogma of performance. The apologia for slowness and hesitation in suboptimality thus echoes the "slow movement"[449]), and its offshoots (slow food, slow tourism, slow science[450]). This movement is even invited in the usually restless video game sector. In particular, in *The Longing*, the hero moves at a snail's pace and has only one goal: to wait for four hundred days![451] Its resonance takes on several forms, highlighting the value of boredom, meditation, naps, holidays, sabbaticals and, simply put, dreaming. In *My Year of Rest and Relaxation*, American writer Ottessa Moshfegh even invented a heroine who decided to put everything aside and hibernate for a full year, because "sleep seemed productive".[452] There is therefore no need for further discussion – the many added values of slowness have already been studied in numerous essays.

It can also be noted in passing that suboptimality is particularly resonant in the Latin world, when it is confronted with the Anglo-Saxon economic efficiency. The slow movement was born out of the protest launched by food critic Carlo Petrini against the opening of a famous American fast-food chain in Rome in 1986. Agroecology really took off in the 1980s in Latin America, in response to the misdeeds of the so-called green revolution. Chilean Miguel Altieri went down in history in this field when he published *Agroecology: The Scientific Basis of Alternative Agriculture* in 1983.[453] Finally, in the footsteps of the Meadows report, Romanian economist and mathematician Nicholas Georgescu-Roegen established the idea of degrowth, which was quickly picked up by Frenchman Jacques Grinevald and then Serge Latouche.

While a long timeframe has its virtues, some sectors are nonetheless less intuitively slow. Could technological innovation, marked by speed and competition, constitute a counter-example? The smartphone, for example, would be an emblematic object of the research conducted in Silicon Valley. In 2015, nearly 2 billion mobile phones were sold and the number of mobile subscribers was surpassing global human population.[454] Innovations of this type demonstrate how Steve Jobs and others had good intuitions. They prove that innovation creates an amplification loop, thus accelerating its own development. Yet when examined more closely, everything that a smartphone is and does, that is, Internet, GPS, touchscreen, batteries, hard disk and voice recognition, are all products of public research, which is in the long run financed by the taxpayer. Although this evidence will not surprise public research players themselves, it appears that the socio-economic world is still living under the illusion that innovation is only accelerated by the market economy. That said, there is no doubt that Steve Jobs was a marketing genius. Thankfully, the real facts finally begin to emerge and showcase the role of time in creation and innovation, notably through the work of Italian economist Mariana Mazzucato.[455] Innovation requires a great deal of time, and often research without immediate application and without short-term benefits.

Time is in fact at the core of the scientific approach, for at least two reasons. First, science has a tendency towards the universal and the comprehensive. As such, it calls for systemic thinking, such as the identification of possible feedback or missing elements relevant to the problem at hand. Such complexity cannot be quickly understood, at least not by a human brain. Second, it is debate and contradiction that establish the robustness of scientific knowledge. As Lebanese poet Khalil Gibran says, "perplexity is the beginning of knowledge". Put another way, confusion is an indispensable step to understanding.

Common sense[396] is not the result of a consensus smoothing out dissonance, but rather of a problematisation that builds on antagonisms. We find here the importance of keeping a strong focus on the issues (promoted by the artistic approach), and therefore the need to devote sufficient time to them, before embarking on the development of solutions.

Thus, science first builds on doubt. As Isaac Asimov puts it "the most exciting phrase to hear in science, the one that heralds new discoveries, is not *Eureka!* but *That's funny . . .*". Conversely, it is the absence of doubt that renders opinions so difficult to reconcile. On a fundamental level, scientific knowledge is born of fruitful conflict, and cannot be reduced to a collection of results. The scientific approach, as a systemic movement, and as an initiatory pathway by way of contradiction, makes use of the virtues of heterogeneity and incoherence over long periods.

Such a slow approach, or a *complex thought*, as Edgar Morin would say,[456] is put into practice outside of the scientific field, in law or politics, for example. One should indeed have enough time to doubt, to develop a debate and to imagine different ways to react. This is what makes slowness a catalyst for suboptimality. One almost has to think backwards! Taking the example cited in the Meadows report, when your car's windshield becomes foggy, you slow down to better perceive your surroundings. Choosing to accelerate would be suicidal, even for "a perfect car and a faultless driver".[37] As we enter the era of uncertainties, slowing down our activities is therefore necessary not only to reduce our impacts on the planet but also to make more room for interactions, to confront the uncertainty to come through more hesitation, and to promote a wider perception of the world.

In the footsteps of Edgar Morin, professor of management David Vallat writes, "learning to decide in a situation of uncertainty involves changing our ways of thinking, to adopt a complex thought, the premise of which is simple: our only certainty is the persistence of uncertainty".[457] Going through the Anthropocene full of certainties would be madness. This again echoes Ashby's law of requisite varieties, or how to confront uncertainty by more uncertainties while allowing, thanks to the long time, a diversity of possible responses to flourish.[307]

Ultimately, time is simply necessary in order to communicate. As philosopher Emmanuel Levinas put it, "time is the other".[458] Yet within a society that is in a process of acceleration, we are confined to immediacy, which prevents mediation and even leads to disinformation. For instance, it has been estimated that in the year 1900 a manager had 22 times longer to make a decision than today.[459] Decisions are thus becoming increasingly intuitive and emotional. Given that our brain also has a tendency to focus on the short term, it is urgent that we slow down in order to be able to communicate, educate and understand the complexity of our world.

A simple path to get us started on our journey. Isabelle Queval, a former elite tennis player and now a university professor and philosopher, acquired the conviction that "the right degree of effort is fulfilment; finding oneself rather than perpetually seeking to outdo oneself".[460] Thus, rather than recruiting generations of pupils to competitive sports classes promoting the dogma of speed and perfectionism, entirely in keeping with the performance society, one could instead support recreational sport or propose the alternative of yoga in school, as a physical means of practicing slowness and robustness. As Ioan Negrutiu, biologist, agronomist and founder of the Michel Serres Institute, puts it, "we often ask ourselves what planet we will leave behind for future generations; we must also ask ourselves which future generations we will leave behind for the planet".

6.22 SERENDIPITY

If time enables reflection, allowing us to communicate and to make better decisions, then this raises the question of choice. Yet humans have difficulty thinking in terms of probability.[394] Often our decisions can be summed up as determinist binary choices, "yes" or "no", and more rarely as "maybes" weighing up the probability of success.

Suboptimality highlights the added value of randomness, in terms of the adaptability that it enables in the long term, at the cost of inefficiency in the short term. How can randomness be better incorporated into our decision-making?

Certain experiences already invite us to wander further. For instance, when you search for a piece of information on Google, you obtain a list of responses that depends on your user profile: the results are ranked based on your previous digital activity. In order to find responses that are further detached from your profile, you will have to access the results further down the offered list. This response is indeed optimised because it is personalised. The search engine Qwant offers far less optimised responses: it does not track your online activities, for ethical reasons, and thus provides a non-biased list of results. This implies that you will probably be more frequently exposed to unexpected, or even undesired results.

What could be the value in a piece of information that is, to whatever degree, unexpected? Because they are founded on advanced personalisation, social networks still provide the same type of information to the same people. These amplification loops generate a form of over-optimisation, ranging from sectarian recruitment to the manipulation of the aforementioned elections. In incorporating randomness to a certain degree, suboptimality could enable the breaking of these cycles. In fact, contingencies such as this would expose users to unsuspected, or even disconcerting information, thus opening up the field of possibility. Likewise, if you search for an article online using keywords, you are going to end up reading articles on the subject you have chosen. If you read the same article in a printed newspaper, you may be tempted to read the neighbouring articles, by chance. This apparent waste of time might facilitate an original response at your next job interview.

The study of the heuristic, the art of making discoveries, can enlighten us.[461] A discovery is very often the result of a combination of pre-existing ideas. We all have a past and a variety of images in our heads. The manner in which we will address an issue and provide a solution to a problem will, to a large extent, depend on this cognitive and cultural baggage. Here past chance encounters will enable the collision of new ideas, provided that the relevance is realised. This is serendipity.

However, the use of randomness in decision-making remains limited, precisely because we all experience the cognitive bias of the status quo that instead favours past experience. When a decision must be made, we very rarely make the deliberate choice to forget all of our preconceived ideas. Justice goes even one step further: jurisprudence is based specifically on the documented knowledge of a past experience. In our everyday life, we do not consider each new problem to be a new, uncharted experience. This would amount to a truly probabilist approach. This very strong cognitive bias can explain, at least in part, why behaving as statisticians – assessing a situation with complete neutrality and offering measured responses – is by no means the norm.

Incidentally, this is the objection that psychologist Daniel Kahneman raises to mathematician Daniel Bernoulli. In his *Utility theory*, Bernoulli established the logic of choice without taking into account human error.[462] The same logic is at work in Kahneman, however, the element of human cognitive bias is incorporated.[88] Among these biases, one might list the illusion of control that drives us to take more risk than is necessary, overestimating the benefits and underestimating the costs. Incidentally, this is a situation that is familiar on many construction sites when it comes to deadlines and final invoices.

Of greater interest to the issue of the Anthropocene is the fact that we have another cognitive bias that moves us to favour what is already familiar. We more readily overlook unknowns, even though we know they exist. The drive towards optimisation within our societies is in keeping with this cognitive bias: we prefer to latch onto what we know, even if this is reduced to a single and therefore fragile solution. For example, we are developing smart cities in spite of increased uncertainty regarding the future of the material resources used to construct digital tools. Similarly, the Covid-19 pandemic has revealed how our over-optimised societies have created a substrate for the development of the pandemic (hyper-mobility, obesity, diabetes etc.). Conversely, the solutions put in place were rather medieval on a global scale (quarantine, masks, distancing, closing borders). In other words, the Covid-19 pandemic did not highlight the virtues of progress (apart from the new technology of mRNA vaccines), but rather revealed how fragile our world of performance and control was becoming, notably through globalised and centralised outsourcing (Chinese masks, Indian vaccines, Taiwanese microprocessors etc.) and by the negative externalities of our behaviours (comorbidities due to diet and stress, hyper-mobility).

At least, suboptimality has the merit of shedding light on the values of the unknown. Counterintuitively, embracing uncertainty before performance is a better guarantee for our future survival. Not that technological advances are useless. Rather, it is the compass of that form of progress that needs to be questioned. If uncertainty and socio-ecological fluctuations becomes the cardinal value of the 21st century, then progress will inevitably shift from performance to robustness, including for technology. The suboptimality of life constitutes a library of paths for managing uncertainties and for building new emergent properties such as autonomy and adaptability. So perhaps we can offer a suggestion to managers who no longer have time for decision-making, and, more broadly, to everyone who might be interested: serendipity is perhaps the first step towards deceleration, and a form of wisdom!

6.23 TO BE CONTINUED

"I always tell the truth: not all of it, because you can't tell all of it". These words of Lacan are borrowed by the writer, philosopher and rabbi Delphine Horvilleur to evoke the value of incompleteness since "truth is in excess. Once said, some remains".

Incompleteness is even a cardinal value of many cultures. The conciseness of Japanese haiku asserts its non-completion to better reflect the evanescence of the world and to arouse the reader's emotion. For example, in this famous haiku by 17th-century Japanese poet Matsuo Bashō:

> "Peace from the old pond,
> A frog jumps in,
> Sound of water".

More broadly, in the artistic sphere, sketches create spaces for interaction, research, imagination, accidents. Incompleteness is a value that can even become tradition, for example, in Judaism: "To live in a house, you must always leave a piece of wall incomplete, or a stone missing from the building [. . .] We are wary of everything that makes us 'one', of everything that makes us complete, of everything that is whole, of everything that defines itself as 'finished' and that prevents the infinite from living in the world and in us. It is as if we must always leave incompleteness to the work".[463] Finally, incompleteness allows the realisation of a trajectory. Early Islamic philosopher Abu Nasr Muhammad al-Farabi (أبو نصر محمد الفارابي) considered this to be one of the fundamental essences of philosophy. He posited that an individual alone might not achieve all aspirations without the support of others. It is a fundamental characteristic of every person to seek interaction with others in the tasks they undertake.[464] This also echoes the word of philosopher Claude Obadia who sees "human as an unfinished being, that is to say, as a bundle of possibilities yet to be realised by them in order to be fulfilled, as if their incompleteness were the price of their freedom".[465]

Having defended the added value of randomness, heterogeneity and incompleteness, it would be paradoxical to finish this essay with a set list of actions, or even a conclusion. This is especially true given that the application of a dose of biological suboptimality to social systems remains, at the very least, debatable. An incomplete counter-model is rather an invitation because it is still "under construction". Thus, shall we attempt a suboptimal conclusion? If we are already in the process of inventing the post-Anthropocene, in which humans coexist with their planet, a "Gaïacene" as it were, then it goes without saying that there are multiple possible solutions. One of the most attractive of these, for me, for you, in this precise moment, is possibly . . .

Some Chronological Reference Points

4.56 billion years ago: Birth of the Earth
4.1 billion years ago: First living organisms (isotopic traces)
3.8 billion years ago: First morphological traces of living organisms (still debated)
3.5 billion years ago: First cell fossils
2.4 billion years ago: Great Oxidation Event
2.1 billion years ago: First multicellular organisms (still debated)
700 million years ago: "Snowball Earth" (global ice age)
445 million years ago: First Phanerozoic mass extinction (end of the Ordovician)
370 million years ago: Second Phanerozoic mass extinction (end of the Devonian)
252 million years ago: Third Phanerozoic mass extinction (end of the Permian, the largest)
201 million years ago: Fourth Phanerozoic mass extinction (end of the Triassic)
66 million years ago: Fifth Phanerozoic mass extinction (end of the Cretaceous)
55 million years ago: Paleocene-Eocene Thermal Maximum
15 million years ago: Heat peak of the Miocene
3.3 million years ago: Start of the Palaeolithic, first stone tools
2.58 million years ago: Start of the Pleistocene (period of glaciation cycles – glaciations actually start 15 million years ago in the Antarctic); start of the human line (*Homo habilis*)
1 million years ago: First traces of mastery of fire (Wonderwerk Cave, South Africa)
315,000 years ago: First *Homo sapiens*
126,000 years ago: Heat peak of the Eemien (penultimate interglacial period of the Pleistocene)
50,000 years ago: First cave paintings (Borneo, Indonesia)
45,000 years ago: Elimination of the Australian megafauna by humans
43,000 years ago: First recorded mine (Ngwenya, Swaziland; hematite for ochre)
29,000 years ago: First recorded pottery (Czech Republic)
11,700 years ago: End of the final period of glaciation (Würm); end of the Pleistocene and the Palaeolithic; start of the Holocene and the Mesolithic
10,500 years ago: First Neolithic revolutions, emergence of agriculture within the Fertile Crescent
6,000 years ago (4000 BC): First cities (Mesopotamia)
5,500 years ago (3500 BC): Invention of writing (end of Prehistory, Mesopotamia); invention of the wheel, irrigation and seed drill (Mesopotamia)
4,500 years ago (2500 BC): First alloy: bronze (copper and tin)
3,000 years ago (1000 BC): High temperature steel industry (1535°C for iron)
8 BC: First known message on paper (flax fibres, China)
From 0 to 500: First global pollution (lead) by Romans

700–1100: Typography (prelude to letterpress printing, China), first paper money (China)
900: Invention of gunpowder (China)
1000: First vaccination (against smallpox, China)
1450: Invention/improvement of the letterpress printing (Johannes Gutenberg)
1492: (Re)discovery of the American continent; start of Columbian trade (sharp increase in the number of domesticated plants and animals, and of infectious diseases, between the ancient and the New World)
1700: Start of the extinctions caused by White westerners (e.g. dodo)
1728: Invention of the punch card for automating looms
1751: First edition of the Encyclopaedia of Diderot and d'Alembert
1776: First steam engines designed by Watt; First Industrial Revolution (coal)
1778: Buffon defines the Anthropocene "The entire face of the Earth today bears the imprint of human power" (Epochs of Nature)
1820: Discovery of disinfecting properties of bleach (Antoine Germain Labarraque)
1825: First railway line opened for the transportation of passengers (United Kingdom)
1858: Laying of the first Atlantic cable (between Ireland and Newfoundland)
1859: First oil well (Pennsylvania, the United States)
1880: Second Industrial Revolution (metallurgy, electricity, petrochemistry);
1882: First implementation of Taylorism
1884–1885: Establishment of the official rules of colonisation at the West Africa Conference (Berlin, Germany)
1891: First high-voltage power line (Germany)
1914–1918: First World War (18.6 million premature deaths)
1918: Fritz Haber is awarded the Nobel Prize in chemistry (synthesis of ammonia)
1927: First transatlantic flight (New York–Paris)
1928: First antibiotic (penicillin, by Alexander Fleming)
1929: American and global market crash
1930: Dust bowl (the United States)
1939–1945: Second World War (60 million premature deaths); start of the military, then civilian, nuclear industry
1943: Decolonisation begins
1946: First computer in the world (ENIAC, University of Pennsylvania)
1950: First theorisation of artificial intelligence (Alan Turing)
1952: Great Smog of London (United Kingdom, 12,000 premature deaths)
1957: Sputnik, the first artificial satellite (USSR)
1959: Demonstration of fishermen against the massive mercury pollution in Minamata (Japan, 900–10,000 premature deaths)
1960s: Green revolution
1969: First humans on the moon (the United States); Early version of the Internet (Advanced Research Projects Agency Network, MIT, Boston, the United States); Lake Erie (the United States) is considered ecologically dead (since revived)
1971: Digital revolution: Invention of the microprocessor, remote networking of around 20 computers, sending of the first email
1972: Meadows report to the Club of Rome; first world environment conference (Stockholm)

1976: Seveso chemical catastrophe (Italy, contamination of 358 ha)
1978: Amoco Cadiz oil spill (France, 227,000 tonnes of crude oil spilled)
1984: Bhopal chemical catastrophe (India)
1986: Chernobyl nuclear catastrophe (Ukraine)
1987: Montreal Protocol to protect the ozone layer
1988: Creation of the International Panel on Climate Change (IPCC)
1997: Discovery of the "plastic continent" of the Pacific North
2000: Beginning of the popularisation of the term *Anthropocene*; burst of the dot-com bubble
2008: Definition of the rights of Mother Earth (Ecuador, Bolivia)
2009: Elinor Ostrom is awarded the Nobel Memorial Prize in economy (governance of the commons)
2010: Definition of the Human Development Index (UN)
2012: International Year of Cooperatives (UN)
2015: Adoption of the Sustainable Development Objectives Agenda (UN)
2016: Coming into force of the Paris Climate Agreement (resulting from the COP21)
2020: Covid-19 crisis: first global confinement

Acronyms and Abbreviations

BP: British Petroleum
CDC: Centers for Disease Control and Prevention
CIA: Central Intelligence Agency
CRM: Critical raw materials
CSA: Community Supported Agriculture
DDT: Dichlorodiphenyltrichloroethane
EISA: Energy Independence and Security Act
ESA: European Space Agency
ETC: Energy Transitions Commission
FAO: Food and Agriculture Organization of the United Nations
GATT: General Agreement on Tariffs and Trade
GDP: Gross Domestic Product
GMO: Genetically Modified Organism
HDI: Human Development Index
INRAE: The French National Research Institute for Agriculture, Food and Environment
IPCC: Intergovernmental Panel on Climate Change
IQ: Intelligence Quotient
LED: Light Emitting Diode
MIT: Massachusetts Institute of Technology (Boston, the United States)
NASA: National Aeronautics and Space Administration
OECD: Organisation for Economic Co-operation and Development
PET: Polyethylene Terephthalate
SDI: Sustainable Development Index
SUV: Sport Utility Vehicle
TED: Technology, Entertainment and Design
UNEP: United Nations Environment Programme
USSR: Union of Soviet Socialist Republics
WEF: World Economic Forum
WHO: World Health Organization
WTO: World Trade Organization
WWF: World Wildlife Fund
ZAD: "Zone d'Aménagement Différé" (French law 26 July 1962 on zones on development stand-by) turned into "Zone A Défendre" (zone to defend) by activists, thus named Zadists

References

1. Serres, M. *Le Contrat Naturel*. (Editions F. Bourin, Paris, 1990).
2. Crutzen, P. J. & Stoermer, E. F. The 'Anthropocene'. *IGBP Newsletter* 17–18 (2000).
3. Lewis, S. L. & Maslin, M. A. Defining the Anthropocene. *Nature* **519**, 171–180 (2015).
4. Latour, B. La terre est enfin ronde. *Libération* (2007).
5. Latour, B. *Où Atterrir? Comment s'orienter En Politique*. (La Découverte, Paris, 2017).
6. Smil, V. *Making the Modern World: Materials and Dematerialization*. (Wiley, Chichester, West Sussex, United Kingdom, 2014).
7. Barnosky, A. D. *et al.* Introducing the *Scientific Consensus on Maintaining Humanity's Life Support Systems in the 21st Century: Information for Policy Makers. Anthr. Rev.* **1**, 78–109 (2014).
8. Magny, M. *Aux racines de l'anthropocène. Une crise écologique reflet d'une crise de l'homme*. (Bord De l'Eau Editions, Latresne, 2019).
9. Steffen, W., Crutzen, P. J. & McNeill, J. R. The Anthropocene: Are humans now overwhelming the great forces of nature. *AMBIO J. Hum. Environ.* **36**, 614–621 (2007).
10. Steffen, W., Broadgate, W., Deutsch, L., Gaffney, O. & Ludwig, C. The trajectory of the Anthropocene: The great acceleration. *Anthr. Rev.* **2**, 81–98 (2015).
11. Rainforest Action Network. *Banking on Climate Change*. https://www.ran.org/wp-content/uploads/2019/04/BOCC_2019_SUMMARY_vUS-1F.pdf (2019).
12. Coady, D., Parry, I., Le, N.-P. & Shang, B. *Global Fossil Fuel Subsidies Remain Large: An Update Based on Country-level Estimates*. (IMF). https://www.imf.org/-/media/Files/Publications/WP/2019/WPIEA2019089.ashx (2019).
13. *The Economist*. The truth about big oil and climate change. (2019).
14. Skolkovo. *The World's Top Auto Markets in 2030: Emerging Markets Transforming the Global Automotive Industry*. https://iems.skolkovo.ru/en/iems/publications/research-reports/157-2010-05-10/ (2010).
15. Devictor, V. *Nature en crise: penser la biodiversité*. (Seuil, Paris, 2015).
16. Hansen, J. *et al.* Ice melt, sea level rise and superstorms: Evidence from paleoclimate data, climate modeling, and modern observations that 2 °C global warming could be dangerous. *Atmospheric Chem. Phys.* **16**, 3761–3812 (2016).
17. Farnsworth, A. & Stone, E. The last time Earth was this hot hippos lived in Britain (that's 130,000 years ago). *The Conversation*. http://theconversation.com/the-last-time-earth-was-this-hot-hippos-lived-in-britain-thats-130-000-years-ago-53398 (2016).
18. Sosdian, S. M. *et al.* Constraining the evolution of Neogene ocean carbonate chemistry using the boron isotope pH proxy. *Earth Planet. Sci. Lett.* **498**, 362–376 (2018).
19. WWF. *Pollution plastique, à qui la faute*. https://www.wwf.fr/vous-informer/actualites/pollution-plastique-a-qui-la-faute (2019).
20. Isobe, A. *et al.* A multilevel dataset of microplastic abundance in the world's upper ocean and the Laurentian Great Lakes. *Micropl. & Nanopl.* **1**, 16 (2021).
21. Siviter, H. *et al.* Agrochemicals interact synergistically to increase bee mortality. *Nature* **596**, 389–392 (2021).
22. FAO. *Évaluation des ressources forestières mondiales*. https://www.fao.org/forest-resources-assessment/past-assessments/fra-2015/fr/ (2015).
23. WWF. *Living Amazon Report 2016*. https://www.worldwildlife.org/publications/living-amazon-report-2016 (2016).
24. Sánchez-Bayo, F. & Wyckhuys, K. A. G. Worldwide decline of the entomofauna: A review of its drivers. *Biol. Conserv.* **232**, 8–27 (2019).

25. Hublin, J.-J. *et al.* New fossils from Jebel Irhoud, Morocco and the pan-African origin of Homo sapiens. *Nature* **546**, 289–292 (2017).
26. Richter, D. *et al.* The age of the hominin fossils from Jebel Irhoud, Morocco, and the origins of the Middle Stone Age. *Nature* **546**, 293–296 (2017).
27. Leakey, R. E. & Lewin, R. *The Sixth Extinction: Biodiversity and Its Survival.* (Phoenix, London, 1999).
28. Barnosky, A. D. *et al.* Has the Earth's sixth mass extinction already arrived? *Nature* **471**, 51–57 (2011).
29. IUCN (International Union for Conservation of Nature). *More Than 27,000 Species Are Threatened with Extinction.* https://www.iucnredlist.org/about (2019).
30. Wilson, E. O. *Biodiversity.* (National Academies Press, Washington, DC, 1988). https://doi.org/10.17226/989.
31. Hammer, K. & Khoshbakht, K. Towards a 'red list' for crop plant species. *Genet. Resour. Crop Evol.* **52**, 249–265 (2005).
32. Vidal, J. 'Tip of the iceberg': Is our destruction of nature responsible for Covid-19? *The Guardian* (2020).
33. Costanza, R. *et al.* Changes in the global value of ecosystem services. *Glob. Environ. Change* **26**, 152–158 (2014).
34. Williams, C. Photos capture the startling effect of shrinking bee populations. *HuffPost* (2016).
35. Bonneuil, C. & Fressoz, J.-B. *L'événement anthropocène: la Terre, l'histoire et nous.* (Éditions Points, Paris, 2016).
36. Meadows, D. H., Randers, J., Meadows, D. L. & Behrens, W. W. *The Limits to Growth: A Report for the Club of Rome's Project on the Predicament of Mankind.* (Universe Books, New York, 1972).
37. Meadows, D. H., Randers, J. & Meadows, D. L. *The Limits to Growth: The 30-Year Update.* (Earthscan, London, 2009).
38. Turner, G. M. On the cusp of global collapse? Updated comparison of *The Limits to Growth* with historical data. *GAIA – Ecol. Perspect. Sci. Soc.* **21**, 116–124 (2012).
39. Rockström, J. *et al.* A safe operating space for humanity. *Nature* **461**, 472–475 (2009).
40. Steffen, W. *et al.* Planetary boundaries: Guiding human development on a changing planet. *Science* **347**, 1259855–1259855 (2015).
41. Richardson, K. *et al.* Earth beyond six of nine planetary boundaries. *Sci. Adv.* **9**, eadh2458 (2023).
42. Steffen, W. *et al.* Trajectories of the earth system in the Anthropocene. *Proc. Natl. Acad. Sci.* **115**, 8252–8259 (2018).
43. Biskaborn, B. K. *et al.* Permafrost is warming at a global scale. *Nat. Commun.* **10**, 264 (2019).
44. AllEnvi. *ScenEnvi: Visions du futur et environnement.* https://www.allenvi.fr/2016-quels-futurs-pour-la-planete-les-scenarii-de-scenenvi/ (2016).
45. Smil, V. *Harvesting the Biosphere: What We Have Taken from Nature.* (MIT Press, Cambridge, 2013).
46. WWF. À partir du 10 mai 2019, l'Union européenne entrera en déficit écologique. *WWF France.* https://www.wwf.fr/vous-informer/actualites/a-partir-du-10-mai-2019-lunion-europeenne-entrera-en-deficit-ecologique (2019).
47. Barnosky, A. D. *et al.* Approaching a state shift in Earth's biosphere. *Nature* **486**, 52–58 (2012).
48. *Le Monde.* A 10 927 mètres au fond du Pacifique, un sac plastique et des emballages de bonbons. (2019).
49. Clark, S. Space debris must be removed from orbit says ESA | Science. *The Guardian.* https://www.theguardian.com/science/across-the-universe/2017/apr/21/space-debris-must-be-removed-from-orbit-says-european-space-agency (2017).

50. *Space Debris by the Numbers*. https://www.esa.int/Space_Safety/Space_Debris/Space_debris_by_the_numbers.
51. Harris, S. Number of children given drugs for ADHD up ninefold with patients as young as THREE being prescribed Ritalin. *Mail Online*. https://www.dailymail.co.uk/health/article-2351427/Number-children-given-drugs-ADHD-ninefold-patients-young-THREE-prescribed-Ritalin.html (2013).
52. Gide, A. *Les faux-monnayeurs*. (Gallimard, Paris, 1925).
53. Spinoza, B. de & Appuhn, C. *Oeuvres 3: éthique*. (Flammarion, Paris, 2002).
54. Harari, Y. N. *Homo Deus: A Brief History of Tomorrow*. (Harvill Secker, London, 2016).
55. Calame, M. *Comprendre l'agroécologie: origines, principes et politiques*. (Éditions Charles Léopold Mayer, Paris, 2016).
56. Smil, V. *Feeding the World: A Challenge for the Twenty-First Century*. (MIT Press, Boston, MA, 2002).
57. Erisman, J. W., Sutton, M. A., Galloway, J., Klimont, Z. & Winiwarter, W. How a century of ammonia synthesis changed the world. *Nat. Geosci.* **1**, 636–639 (2008).
58. Ritchie, H. *How many people does synthetic fertilizer feed*? https://ourworldindata.org/how-many-people-does-synthetic-fertilizer-feed (2017).
59. Hamilton, C. *Les apprentis sorciers du climat: raisons et déraisons de la géo-ingénierie*. (Seuil, Paris, 2013).
60. Smith, P. *et al.* Biophysical and economic limits to negative CO2 emissions. *Nat. Clim. Change* **6**, 42–50 (2016).
61. Heck, V., Gerten, D., Lucht, W. & Popp, A. Biomass-based negative emissions difficult to reconcile with planetary boundaries. *Nat. Clim. Change* **8**, 151–155 (2018).
62. Smetacek, V. *et al.* Deep carbon export from a Southern Ocean iron-fertilized diatom bloom. *Nature* **487**, 313–319 (2012).
63. Deluzarche, C. Géo-ingénierie: ensemencer l'océan de fer n'aidera pas à limiter le réchauffement climatique. *Futura*. https://www.futura-sciences.com/planete/actualites/ocean-geo-ingenierie-ensemencer-ocean-fer-naidera-pas-limiter-rechauffement-climatique-56039/ (2020).
64. Crutzen, P. J. Albedo Enhancement by stratospheric sulfur injections: A contribution to resolve a policy dilemma? *Clim. Change* **77**, 211–220 (2006).
65. Watts, J. US and Saudi Arabia blocking regulation of geoengineering, sources say. *The Guardian* (2019).
66. Bregman, R. Friedman and Hayek: How neoliberal revolutionaries took over the world. *Evonomics* (2016).
67. Horkheimer, M., Adorno, T. W., Kaufholz-Messmer, É. & Horkheimer, M. *La dialectique de la raison: Fragments philosophiquies*. (Gallimard, s.l., 1974).
68. Dorst, J. *La force du vivant*. (Flammarion, Paris, 1981).
69. Diamond, J. M. *Collapse: How Societies Choose to Fail or Succeed*. (Penguin Books, New York, 2011).
70. Costanzo, E. & Hamant, O. L'Anthropocène à la lumière des rétroactions biologiques. In *Anthropocène à l'école de l'indiscipline* (Temps circulaire, Lyon, 2016).
71. Rattinacannou, J.-E. Après la surpêche, la morue se refait une santé. *Futura* (2011).
72. Lustgarten, A. palm oil was supposed to help save the planet. Instead it unleashed a catastrophe. *The New York Times* (2018).
73. Searchinger, T. *et al.* Use of U.S. croplands for biofuels increases greenhouse gases through emissions from land-use change. *Science* **319**, 1238–1240 (2008).
74. Kadandale, S., Marten, R. & Smith, R. The palm oil industry and noncommunicable diseases. *Bull. World Health Organ.* **97**, 118–128 (2019).
75. Phoenix, M., Kroff, F. & Eggen, M. *Land Grabbing for Palm Oil in Sierra Leone: Case Report 2019*. https://www.fian.be/IMG/pdf/fian_b_report_landgrab_in_sl_malen_2019_full_weblow.pdf (2019).

76. Berrod, N. Incendies en Amazonie: quelle est la "part de complicité" de la France? *leparisien.fr* (2019).
77. Carson, R., Darling, L. & Darling, L. *Silent Spring*. (Houghton Mifflin; Riverside Press, Boston; Cambridge, MA, 1962).
78. O'Shaughnessy, P. *Operation Cat Drop*. http://catdrop.com/ (2019).
79. Storeng, K. T. The GAVI alliance and the 'Gates approach' to health system strengthening. *Glob. Public Health* **9**, 865–879 (2014).
80. The Lancet. What has the Gates Foundation done for global health? *The Lancet* **373**, 1577 (2009).
81. Jensen, P. *Pourquoi La Société Ne Se Laisse Pas Mettre En Équations*. (Éditions du Seuil, Paris, 2018).
82. FAO. *La situation mondiale des pêches et de l'aquaculture*. http://www.fao.org/3/a-i5555f.pdf (2016).
83. Finley, C. *All the Fish in the Sea: Maximum Sustainable Yield and the Failure of Fisheries Management*. (University of Chicago Press, Chicago; London, 2011).
84. Kübler-Ross, E., Kessler, D. & Touati, J. *Sur le chagrin et sur le deuil: trouver un sens à sa peine à travers les cinq étapes du deuil*. (Pocket, Paris, 2011).
85. McCarthy, N. Oil and gas giants spend millions lobbying to block climate change policies [Infographic]. *Forbes*. https://www.forbes.com/sites/niallmccarthy/2019/03/25/oil-and-gas-giants-spend-millions-lobbying-to-block-climate-change-policies-infographic/ (2019).
86. Arendt, H. *Eichmann à Jerusalem: Rapport Sur La Banalité Du Mal*. (Gallimard, Paris, 1991).
87. Milgram, S. Behavioral study of obedience. *J. Abnorm. Soc. Psychol.* **67**, 371–378 (1963).
88. Kahneman, D. *Thinking, Fast and Slow*. (Farrar, Straus and Giroux, New York, 2011).
89. Festinger, L. *A Theory of Cognitive Dissonance*. (Stanford University Press, Stanford, 1957).
90. Festinger, L., Schachter, S. & Riecken, H. W. *When Prophecy Fails*. (Pinter Martin, London, 2009).
91. Gaines, B. J., Kuklinski, J. H., Quirk, P. J., Peyton, B. & Verkuilen, J. Same facts, different interpretations: Partisan motivation and opinion on Iraq. *J. Polit.* **69**, 957–974 (2007).
92. Klein, E. How politics makes us stupid. *Vox* (2014).
93. Kahan, D. M., Peters, E., Dawson, E. C. & Slovic, P. *Motivated numeracy and enlightened self-government. Behav. Public Policy* **1**, 54–86 (2013).
94. Kuhn, T. S. *The Structure of Scientific Revolutions*. (University of Chicago Press, Chicago, IL, 1996).
95. Rifkin, J. *The European Dream: How Europe's Vision of the Future Is Quietly Eclipsing the American Dream*. (Jeremy P. Tarcher/Penguin, New York, NY, 2004).
96. Jevons, W. S. *The Coal Question: An Inquiry Concerning the Progress of the Nation, and the Probable Exhaustion of Our Coal-Mines*. (Macmillan & Co, London; Cambridge, UK, 1865).
97. Dennis, M. A. Drilling for dollars: The making of US petroleum reserve estimates, 1921–25. *Soc. Stud. Sci.* **15**, 241–265 (1985).
98. Ickes, H. War and our vanishing resources. *American Magazine* 18–23 (1945).
99. IPCC. *CLIMATE CHANGE 2013: The Physical Science Basis*. https://www.ipcc.ch/site/assets/uploads/2017/09/WG1AR5_Frontmatter_FINAL.pdf (2013).
100. Robson, D. *The '3.5% Rule': How a Small Minority Can Change the World*. https://www.bbc.com/future/article/20190513-it-only-takes-35-of-people-to-change-the-world (2019).
101. Davis, H. & Turpin, E. *Art in the Anthropocene: Encounters among Aesthetics, Politics, Environments and Epistemologies 2015*. (Open Humanities Press, London, 2015).

102. Logé, G. *Renaissance sauvage. L'art de l'anthropocène*. (PUF – HUMENSIS, Paris, 2019).
103. Lewallen, C., Seid, S. & Lord, C. *Ant Farm, 1968–1978*. (University of California Press: Berkeley Art Museum: Pacific Film Archive, Berkeley, 2004).
104. Francis, M. *Les années pop: 1956–1968; exposition présentées au Centre Pompidou, Galerie 1, 15 mars – 18 juin 2001*. (Centre Pompidou, Paris, 2001).
105. Burke, E. *Recherche Philosophique Sur l'origine de Nos Idées Du Sublime et Du Beau*. (Pichon, Paris, 1803).
106. Lucretius, T. C. & Ernout, A. *De la nature*. (Les Belles Lettres, Paris, 2019).
107. Blanc, N. De l'invention du paysage à l'ère de l'anthropocène. In *Anthropocène à l'école de l'indiscipline* (Temps circulaire, Lyon, 2016).
108. Descola, P. *Les Natures En Question: Colloque Annuel 2017*. (Odile Jacob, Paris, 2018).
109. Brundtland, G. *Our Common Future. Report of the World Commission on Environment and Development*. https://sustainabledevelopment.un.org/content/documents/5987our-common-future.pdf (1987).
110. Robinson, J. Squaring the circle? Some thoughts on the idea of sustainable development. *Ecol. Econ.* **48**, 369–384 (2004).
111. ETC. *Better Energy, Greater Prosperity Achievable Pathways to Low-carbon Energy Systems*. https://www.energy-transitions.org/publications/better-energy-greater-prosperity/ (2017).
112. Mitcham, C. The concept of sustainable development: Its origins and ambivalence. *Technol. Soc.* **17**, 311–326 (1995).
113. Daly, H. E. *Beyond Growth: The Economics of Sustainable Development*. (Beacon Press, Boston, Mass, 1996).
114. Geyer, R., Jambeck, J. R. & Law, K. L. Production, use, and fate of all plastics ever made. *Sci. Adv.* **3**, e1700782 (2017).
115. McCormick, E. *et al. Where Does Your Plastic Go? Global Investigation Reveals America's Dirty Secret Erin McCormick, Bennett Murray, Carmela Fonbuena, Leonie Kijewski, Gökçe, Jamie Fullerton, Alastair Gee and Charlotte Simmonds*. https://www.theguardian.com/us-news/2019/jun/17/recycled-plastic-america-global-crisis (2019).
116. Zero Waste Cities. Zero waste cities. *Zero Waste Cities*. https://zerowastecities.eu/ (2019).
117. Ademe. *Batiactu 'L'objectif 2020 est possible'*. https://www.batiactu.com/edito/batiments-a-energie-positive-objectif-2020-est-possible-32463.php (2012).
118. Hopkins, R. *Ils changent le monde!: 1.001 initiatives de transition écologique*. (Seuil, Paris, 2014).
119. Dion, C. & Laurent, M. *Demain*. (Mars Distribution, Paris, 2015).
120. PNUE. *Rapport du Panel international des ressources du Programme des Nations unies pour l'environnement*. https://www.un.org/youthenvoy/fr/2013/08/pnue-programme-nations-unies-lenvironnement/ (2013).
121. Tamara, S., Hoek, M., Scheltema, R. A., Leney, A. C. & Heck, A. J. R. A colorful pallet of b-phycoerythrin proteoforms exposed by a multimodal mass spectrometry approach. *Chem* **5**, 1302–1317 (2019).
122. Pelamis. Pelamis. *L'énergie des vagues*. http://wavepower.ek.la/pelamis-p488823 (2008).
123. Bihouix, P. *L'âge Des Low Tech: Vers Une Civilisation Techniquement Soutenable*. (Éditions du Seuil, Paris, 2014).
124. Assouly, O. *et al. Manifeste pour une exception agricole et écologique*. https://alimentation-generale.fr/chronique/pour-une-exception-agricole-et-ecologique/ (2017).
125. Collart Dutilleul, F. *La Charte de La Havane: pour une autre mondialisation*. (Dalloz, Paris, 2018).
126. Smil, V. *BP Statistical Review of World Energy*. https://www.bp.com/content/dam/bp/business-sites/en/global/corporate/pdfs/energy-economics/statistical-review/bp-statistical-review-of-world-energy-2017-full-report.pdf (2017).

127. Hickel, J. Degrowth: A theory of radical abundance. *Real-world Economics Review* 54–68 (2019).
128. Foster, J. B., Clark, B. & York, R. Capitalism and the curse of energy efficiency: The return of the Jevons Paradox. *Mon. Rev.* **62**, 1 (2010).
129. Erhardt, G. D. *et al.* Do transportation network companies decrease or increase congestion? *Sci. Adv.* **5**, eaau2670 (2019).
130. Grossman, G. & Krueger, A. *Environmental Impacts of a North American Free Trade Agreement.* http://www.nber.org/papers/w3914.pdf (1991). https://doi.org/10.3386/w3914.
131. Meunier, A. *Controverses autour de la courbe environnementale de Kuznets.* Documents de travail 107, Groupe d'Economie du Développement de l'Université Montesquieu Bordeaux IV (2004).
132. Beuret, N. *Emissions Inequality: There Is a Gulf between Global Rich and Poor.* https://wid.world/news-article/climate-change-the-global-inequality-of-carbon-emissions (2019).
133. Felli, R. *La Grande Adaptation: Climat, Capitalisme et Catastrophe.* (Éditions du Seuil, Paris, 2016).
134. GMO answers. *Do GMOs Lead to Increased Pesticide Use?* https://gmoanswers.com/do-gmos-lead-increased-pesticide-use (2018).
135. FAO. *Questionnaire on Pesticides Use and Land.* https://www.fao.org/fileadmin/templates/agphome/documents/Pests_Pesticides/Code/2017_Questionnaire_FAO_WHO/Questionnaire_on_Pesticide_Management__Part_II__Agriculture.docx (2014).
136. Mandart, S. & Foucart, S. Le recours aux pesticides a connu une hausse spectaculaire en 2018. *Le Monde* (2020).
137. Mandart, S. Les chiffres noirs des ventes de pesticides " extrêmement dangereux ". *Le Monde* (2020).
138. Pesticide Action Network UK. *The Hidden Rise of UK Pesticide Use: Fact-checking an Industry Claim.* https://www.pan-uk.org/pesticides-agriculture-uk/ (2018).
139. Robaglia, C. & Caranta, C. Translation initiation factors: A weak link in plant RNA virus infection. *Trends Plant Sci.* **11**, 40–45 (2006).
140. Iglesias, V. A. & Meins, F. Movement of plant viruses is delayed in a beta-1,3-glucanase-deficient mutant showing a reduced plasmodesmatal size exclusion limit and enhanced callose deposition. *Plant J.* **21**, 157–166 (2000).
141. Ecophyto. *Plan Écophyto – Réduire le recours aux produits phytosanitaires | Alim'agri.* https://agriculture.gouv.fr/plan-ecophyto-reduire-le-recours-aux-produits-phytosanitaires (2018).
142. INRA. *Agriculture sans pesticides: l'Inra moteur dans la construction d'une feuille de route européenne.* https://www.inrae.fr/actualites/agriculture-pesticides-inrae-moteur-construction-dune-feuille-route-europeenne (2019).
143. Valantin, J.-M. *Géopolitique d'une Planète Déréglée: Le Choc de l'anthropocène.* (Éditions du Seuil, Paris, 2017).
144. Robert, A. Le marché du carbone renaît de ses cendres. *Reporterre, le quotidien de l'écologie.* https://reporterre.net/Le-marche-du-carbone-renait-de-ses-cendres (2021).
145. Pottier, A. & Giraud, G. *Comment Les Économistes Réchauffent La Planète.* (Éditions du Seuil, Paris, 2016).
146. Huzar, E. *L'arbre de La Science.* (1857).
147. Noghès, Y.-A. Zora, le robot qui fait danser les seniors en maison de retraite. *BFMTV.* https://www.bfmtv.com/mediaplayer/video/zora-le-robot-qui-fait-danser-les-seniors-en-maison-de-retraite-508873.html (2015).
148. Grumbach, S. & Hamant, O. Digital revolution or anthropocenic feedback? *Anthr. Rev.* **5**, 87–96 (2018).

149. Astier, M. Guillaume Pitron: “ Un téléphone portable ne pèse pas 150 grammes, mais 150 kilos ”. *Reporterre, le quotidien de l'écologie.* https://reporterre.net/Guillaume-Pitron-Un-telephone-portable-ne-pese-pas-150-grammes-mais-150-kilos (2021).
150. Cohen, D. Earth's natural wealth: An audit. *New Scientist.* https://www.newscientist.com/article/mg19426051-200-earths-natural-wealth-an-audit/ (2007).
151. ANCRE. *Ressources minérales et énergie.* https://www.allianceenergie.fr/wp-content/uploads/2020/11/AncrePositionPaperRessources.pdf (2015).
152. Auzanneau, M. La production américaine de pétrole a cessé de croître. *Oil Man.* https://www.lemonde.fr/blog/petrole/2019/09/13/la-production-americaine-de-petrole-a-cesse-de-croitre/ (2019).
153. WEF. *WEF Global Agenda Survey.* https://www.weforum.org/publications/global-agenda-survey-2012-report/ (2012).
154. Achzet, B. *et al. Materials Critical to the Energy Industry. An Introduction.* https://www.gdch.de/fileadmin/downloads/Netzwerk_und_Strukturen/Fachgruppen/Vereinigung_fuer_Chemie_und_Wirtschaft/vortraege_2011/achzet.pdf (2011).
155. Krum, R. *The Periodic Table of iPhones.* https://coolinfographics.com/blog/2013/2/4/the-periodic-table-of-iphones.html (2013).
156. Walsh, B. *E-Waste Not How – and Why – We Should Make Sure Our Old Cell Phones, TVs and PCs Get Dismantled Properly By Bryan Walsh Thursday, Jan. 08, 2009.* https://content.time.com/time/magazine/article/0,9171,1870485,00.html (2009).
157. CIA. *CIA World Factbook.* https://www.cia.gov/the-world-factbook/ (Langley, VA, USA, 2015).
158. ECHA (European Chemicals Agency). *Study Report on the Conditions of Use of Five Cobalt Salts – European Chemicals Agency.* https://echa.europa.eu/documents/10162/13641/cobalts_salts_study_report_en.pdf/42f0947f-e7fe-7b14-fc97-cfda0c068e9d (2017).
159. UNEP & GRID-Geneva. *Sand and Sustainability: Finding New Solutions for Environmental Governance of Global Sand Resources: Synthesis for Policy Makers.* https://wedocs.unep.org/handle/20.500.11822/28163 (2019).
160. Combe, M. *La pénurie de sable, c'est pour bientôt!* https://www.natura-sciences.com/comprendre/penurie-sable.html (2018).
161. Gillis, J. R. Why sand is disappearing. *New York Times* (2014).
162. Cordell, D., Drangert, J.-O. & White, S. The story of phosphorus: Global food security and food for thought. *Glob. Environ. Change* **19**, 292–305 (2009).
163. Vaccari, D. Phosphore: une crise imminente. *Pour la science* 36–41 (Belin, Paris, 2010).
164. Blackwell, M., Darch, T. & Haslam, R. Phosphorus use efficiency and fertilizers: Future opportunities for improvements. *Front. Agric. Sci. Eng.* **0**, 0 (2019).
165. Faradji, C. & de Boer, M. *How the Great Phosphorus Shortage Could Leave Us All Hungry.* https://theconversation.com/how-the-great-phosphorus-shortage-could-leave-us-all-hungry-54432 (2016).
166. Carrington, D. Phosphate fertiliser ‘crisis’ threatens world food supply. *The Guardian* (2019).
167. Darch, T. *et al.* Fertilizer produced from abattoir waste can contribute to phosphorus sustainability, and biofortify crops with minerals. *PLOS ONE* **14**, e0221647 (2019).
168. Shi, J. *et al.* A phosphate starvation response-centered network regulates mycorrhizal symbiosis. *Cell* **184**, 5527–5540.e18 (2021).
169. FAO. *L'etat de la securite alimentaire et de la nutrition dans le monde.* https://www.fao.org/publications/home/fao-flagship-publications/the-state-of-food-security-and-nutrition-in-the-world/2022/fr (2019).
170. Cicolella, A. *Toxique planète le scandale invisible des maladies chroniques.* (Seuil, Paris, 2013).
171. Wynes, S. & Nicholas, K. A. The climate mitigation gap: Education and government recommendations miss the most effective individual actions Seth Wynes1,2,3 and Kimberly A Nicholas1. *Environ. Res. Lett.* 074024 (2017).

172. Sengupta, S. & Cai, W. A quarter of humanity faces looming water crises. *The New York Times* (2019).
173. Michaelson, R. Oil built Saudi Arabia – will a lack of water destroy it? *The Guardian* (2019).
174. Robbins, J. As water scarcity increases, desalination plants are on the rise. *Yale E360* (2019).
175. Jones, E., Qadir, M., van Vliet, M. T. H., Smakhtin, V. & Kang, S. The state of desalination and brine production: A global outlook. *Sci. Total Environ.* **657**, 1343–1356 (2019).
176. Padilla, E. Un monde sans eau. Au Honduras, les paysans fuient le corridor de la sécheresse. *Criterio.* https://www.courrierinternational.com/article/un-monde-sans-eau-au-honduras-les-paysans-fuient-le-corridor-de-la-secheresse (2019).
177. Telegraph reporters. England set to run short of water within 25 years, Environment Agency warns. *The Telegraph* (2019).
178. Melissa. *MELISSA: de la recherche spatiale utile sur la Terre.* http://www.esa.int/Space_in_Member_States/Belgium_-_Francais/MELISSA_de_la_recherche_spatiale_utile_sur_la_Terre (2016).
179. IPCC. *Global Warming of 1.5°C.* https://www.ipcc.ch/sr15/ (2018).
180. PNUE. *Single-use Plastics: A Roadmap for Sustainability.* https://www.unep.org/resources/report/single-use-plastics-roadmap-sustainability (2018).
181. JIS. *Jakarta Intercultural School: Fees.* https://www.jisedu.or.id/admissions/fees (2019).
182. Servigne, P. & Stevens, R. *Comment tout peut s'effondrer: petit manuel de collapsologie à l'usage des générations présentes.* (Seuil, Paris, 2015).
183. Charbonneau, B. & Ellul, J. *'Nous Sommes Des Révolutionnaires Malgré Nous': Textes Pionniers de l'écologie Politique.* (Éditions du Seuil, Paris, 2014).
184. Alberti, L. B. *De Pictura.* (1437).
185. Trouslard, Y. L'Anthropocène en perspective. In *Anthropocène à l'école de l'indiscipline* 191–195 (Temps circulaire, Lyon, 2016).
186. Inra. *Document d'orientation #Inra2025.* https://hal.science/hal-01607768/document (2017).
187. Serres, M. *Temps Des Crises.* (Pommier, Paris, 2009).
188. Benoit-Browaeys, D. *L'urgence du vivant: vers une nouvelle économie.* (François Bourin, Paris, 2018).
189. Noisette, C. France: un rapport officiel craint une utilisation malveillante des nouveaux OGM. *Inf'OGM.* https://www.infogm.org/6304-france-cncb-craint-utilisation-malveillante-nouveaux-ogm (2017).
190. Descola, P. *Les lances du crépuscule: relations jivaros, Haute Amazonie.* (Pocket, Paris, 2006).
191. USGS (United States Geological Survey). *Phosphate Rock.* https://www.usgs.gov/centers/national-minerals-information-center/phosphate-rock-statistics-and-information (2009).
192. Buffon, G. L. L. *Les époques de la nature: 1778.* (Paleo, Clermont-Ferrand, 1778).
193. Haraway, D. J. *Staying with the Trouble: Making Kin in the Chthulucene.* (Duke University Press, Durham, 2016).
194. Robinet, J.-B. *De La Nature.* Vol. 4 (Amsterdam, 1776).
195. Nogaret, F. *La Terre Est Un Animal.* (Versailles, 1795).
196. Lovelock, J. E. & Margulis, L. Atmospheric homeostasis by and for the biosphere: The Gaia hypothesis. *Tellus* **26**, 2–10 (1974).
197. Laland, K. N., Odling-Smee, F. J. & Feldman, M. W. Evolutionary consequences of niche construction and their implications for ecology. *Proc. Natl. Acad. Sci.* **96**, 10242–10247 (1999).
198. Jones, C. G., Lawton, J. H. & Shachak, M. Organisms as ecosystem engineers. *Oikos* **69**, 373 (1994).

199. Gilbert, S. F., Sapp, J. & Tauber, A. I. A symbiotic view of life: We have never been individuals. *Q. Rev. Biol.* **87**, 325–341 (2012).
200. Bataille, G. *La part maudite*. (Éditions de Minuit, Paris, 1949).
201. Kendrick, J. W. The historical development of national-income accounts. *Hist. Polit. Econ.* **2**, 284–315 (1970).
202. Jericho, G. If you are under 34, you have never experienced a month of below average temperatures | Greg Jericho | Opinion. *The Guardian*. https://www.theguardian.com/commentisfree/2019/oct/10/if-you-are-under-34-youve-experienced-just-one-month-of-below-average-temperatures (2019).
203. Georgescu-Roegen, N. *The Entropy Law and the Economic Process*. (Harvard University Press, Erscheinungsort nicht ermittelbar, 1971).
204. Davis, M. & Saint-Upéry, M. *Génocides tropicaux: catastrophes naturelles et famines coloniales (1870–1900): aux origines du sous-dévelopement*. (La Découverte, Paris, 2006).
205. PNUE. *Farming Systems Principles for Improved Food Production and the Control of Soil Degradation in the Arid, Semi-arid, and Humid Tropics: Summary Proceedings of an Experts' Meeting*. https://oar.icrisat.org/854/ (1986).
206. Lelieveld, J. *et al.* Cardiovascular disease burden from ambient air pollution in Europe reassessed using novel hazard ratio functions. *Eur. Heart J.* **40**, 1590–1596 (2019).
207. *The Guardian*. Air pollution is the 'new tobacco', warns WHO head. (2018).
208. Escardio. *The World Faces an Air Pollution 'pandemic'*. https://www.escardio.org/The-ESC/Press-Office/Press-releases/The-world-faces-an-air-pollution-pandemic, https://www.escardio.org/The-ESC/Press-Office/Press-releases/The-world-faces-an-air-pollution-pandemic (2020).
209. Lelieveld, J. *et al.* Loss of life expectancy from air pollution compared to other risk factors: A worldwide perspective. *Cardiovasc. Res.* **116**, 1910–1917. https://doi.org/10.1093/cvr/cvaa025 (2020).
210. Steliarova-Foucher, E. *et al.* International incidence of childhood cancer, 2001–10: A population-based registry study. *Lancet Oncol.* **18**, 719–731 (2017).
211. Lainé, N. Pratiques ethno-vétérinaires sur les éléphants au Laos: Un savoir co-construit avec les animaux? *Rev. D'ethnoécologie* **17**. https://doi.org/10.4000/ethnoecologie.5917 (2020).
212. Chartier, M. *Les data centers de Google consomment plus de 16 milliards de litres d'eau par an aux États-Unis*. https://www.lesnumeriques.com/vie-du-net/les-data-centers-de-google-consomment-plus-de-16-milliards-de-litres-d-eau-par-an-aux-etats-unis-n204639.html (2023).
213. Vaughan, A. Cloud gaming may be great for gamers but bad for energy consumption. *New Scientist*. https://www.newscientist.com/article/2206200-cloud-gaming-may-be-great-for-gamers-but-bad-for-energy-consumption/ (2019).
214. Jones, N. How to stop data centres from gobbling up the world's electricity. *Nature* **561**, 163–166 (2018).
215. Malm, A. *Fossil Capital: The Rise of Steam-Power and the Roots of Global Warming*. (Verso, London; New York, 2016).
216. Mitchell, T. *Carbon Democracy: Political Power in the Age of Oil*. (Verso, London, 2013).
217. Beck, U. *La société du risque: sur la voie d'une autre modernité*. (Flammarion, Paris, 2015).
218. Illich, I. *Tools for Conviviality*. (Calder and Boyars, London, 1973).
219. Dupuy, J.-P. *Pour un catastrophisme éclairé: Quand l'impossible est certain*. (Seuil, Paris, 2004).
220. Callahan, P. Amazon pushes fast shipping but avoids responsibility for the human cost. *The New York Times* (2019).

221. Edwards, M. A. & Roy, S. Academic research in the 21st century: Maintaining scientific integrity in a climate of perverse incentives and hypercompetition. *Environ. Eng. Sci.* **34**, 51–61 (2017).
222. Francescani, C. NYPD report confirms manipulation of crime stats. *Reuters* (2012).
223. Campbell, D. T. Assessing the impact of planned social change. *Eval. Program Plann.* **2**, 67–90 (1979).
224. Fire, M. & Guestrin, C. Over-optimization of academic publishing metrics: Observing Goodhart's Law in action. *GigaScience* **8**, giz053 (2019).
225. Fressoz, J.-B. *L'apocalypse Joyeuse: Une Histoire Du Risque Technologique*. (Éditions du Seuil, Paris, 2012).
226. Macnab, R. M. How bacteria assemble flagella. *Annu. Rev. Microbiol.* **57**, 77–100 (2003).
227. Silverman, M. & Simon, M. Flagellar rotation and the mechanism of bacterial motility. *Nature* **249**, 73–74 (1974).
228. Berg, H. C. & Anderson, R. A. Bacteria swim by rotating their flagellar filaments. *Nature* **245**, 380–382 (1973).
229. Darwin, C. *On the Origin of Species by Means of Natural Selection, or the Preservation of Favoured Races in the Struggle for Life*. (1859).
230. Lamarck, J.-B. P. A. de M. de. *Philosophie zoologique: ou Exposition des considérations relatives à l'histoire naturelle des animaux*. (Ed. Flammarion, Paris, 1809).
231. Tsimring, L. S. Noise in biology. *Rep. Prog. Phys. Phys. Soc. G. B.* **77**, 026601 (2014).
232. Raj, A., Rifkin, S. A., Andersen, E. & van Oudenaarden, A. Variability in gene expression underlies incomplete penetrance. *Nature* **463**, 913–918 (2010).
233. Abley, K., Locke, J. C. W. & Leyser, H. M. O. Developmental mechanisms underlying variable, invariant and plastic phenotypes. *Ann. Bot.* **117**, 733–748 (2016).
234. Ewens, W. J. James F. Crow and the stochastic theory of population genetics. *Genetics* **190**, 287–290 (2012).
235. Chapelle, G., Decoust, M., Hulot, N., Pelt, J.-M. & Schuiten, L. *Le Vivant Comme Modèle*. (Albin Michel, Paris, 2015).
236. Adesina, O., Anzai, I. A., Avalos, J. L. & Barstow, B. Embracing biological solutions to the sustainable energy challenge. *Chem* **2**, 20–51 (2017).
237. Blankenship, R. E. *et al.* Comparing photosynthetic and photovoltaic efficiencies and recognizing the potential for improvement. *Science* **332**, 805–809 (2011).
238. Zelenitsky, D. K. *et al.* Feathered non-avian dinosaurs from North America provide insight into wing origins. *Science* **338**, 510–514 (2012).
239. Wolpert, L. Positional information and the spatial pattern of cellular differentiation. *J. Theor. Biol.* **25**, 1–47 (1969).
240. Meinhardt, H. Cell determination boundaries as organizing regions for secondary embryonic fields. *Dev. Biol.* **96**, 375–385 (1983).
241. Blackburn, D. G. & Flemming, A. F. Morphology, development, and evolution of fetal membranes and placentation in squamate reptiles. *J. Exp. Zoolog. B Mol. Dev. Evol.* **312B**, 579–589 (2009).
242. Kelly, W. G. Transgenerational epigenetics in the germline cycle of Caenorhabditis elegans. *Epigenetics Chromatin* **7**, 6 (2014).
243. Heard, E. & Martienssen, R. A. Transgenerational epigenetic inheritance: Myths and mechanisms. *Cell* **157**, 95–109 (2014).
244. Ciabrelli, F. *et al.* Stable Polycomb-dependent transgenerational inheritance of chromatin states in Drosophila. *Nat. Genet.* **49**, 876–886 (2017).
245. Martienssen, R. A. DNA methylation and epigenetic inheritance in plants and filamentous fungi. *Science* **293**, 1070–1074 (2001).
246. Geoghehan, B., Grumbach, S., Halpern, O., Hamant, O. & Mitchell, R. *Algorithmic Intermediation and Smartness*. https://www.anthropocene-curriculum.org/project/technosphere/campus-2016/seminar-algorithmic-intermediation-and-smartness (2016).

247. Liu, Y. *et al.* Detecting cancer metastases on gigapixel pathology images. *ArXiv* (2017).
248. Abbati, G. *et al.* MRI-based surgical planning for lumbar spinal stenosis. In *Medical Image Computing and Computer-Assisted Intervention – MICCAI 2017* (eds. Descoteaux, M. *et al.*), Vol. 10435, 116–124 (Springer International Publishing, Cham, 2017).
249. Dugain, M. & Labbé, C. *L'homme nu. Le livre noir de la révolution numérique.* (Editions Plon, Paris, 2016).
250. Tréguer, F. *L'utopie Déchue: Une Contre-Histoire d'Internet, XVe-XXIe Siècle.* (Fayard, Paris, 2019).
251. Melville, H. *Bartleby le scribe.* (Gallimard, Paris, 1853).
252. Michelet, J. *Tableau de la France.* (Equateurs, Sainte-Marguerite-sur-Mer, 1861).
253. Michelet, J. Leçons de M. Michelet. In *Des Jésuites.* (Hachette, Paris, 1843).
254. Harper, C. V. *et al.* Temperature regulates NF-κB dynamics and function through timing of A20 transcription. *Proc. Natl. Acad. Sci.* **115**, E5243–E5249 (2018).
255. Munda, G. "Measuring Sustainability": A multi-criterion framework. *Environ. Dev. Sustain.* **7**, 117–134 (2005).
256. Bernard, C. *La Théorie de La Viabilité Au Service de La Modélisation Mathématique Du Développement Durable. Application Au Cas de La Forêt Humide de Madagascar.* (Université Blaise Pascal, Clermont-Ferrand, 2011).
257. Aubin, J. P. *Viability Theory.* (Birkhäuser, Boston, 2009).
258. Doyen, L. & Saint-Pierre, P. Scale of viability and minimal time of crisis. *Set-Value Anal.* 227–246 (1997).
259. Bonneuil, N. & Saint-Pierre, P. Beyond optimality: Managing children, assets, and consumption over the life cycle. *J. Math. Econ.* **44**, 227–241 (2008).
260. Bonneuil, N. & Saint-Pierre, P. Protected polymorphism in the two-locus haploid model with unpredictable fitnesses. *J. Math. Biol.* **40**, 251–277 (2000).
261. Andres-Domenech, P., Saint-Pierre, P. & Zarzour, G. Forest conservation and co2 emissions: A viable approach. *Les cahiers du GERAD* 25 (2008).
262. Martinet, V., Thébaud, O. & Doyen, L. Defining viable recovery paths toward sustainable fisheries. *Ecol. Econ.* **64**, 411–422 (2007).
263. Hardin, G. The competitive exclusion principle. *Science* **131**, 1292–1297 (1960).
264. Pennac, D. *Comme Un Roman.* (Gallimard, Paris, 1992).
265. Ribault, T. *Contre la résilience: à Fukushima et ailleurs.* (l'Échappée, Paris, 2021).
266. Folke, C. *et al.* Resilience thinking: Integrating resilience, adaptability and transformability. *Ecol. Soc.* **15**, (2010).
267. Penn, J. L., Deutsch, C., Payne, J. L. & Sperling, E. A. Temperature-dependent hypoxia explains biogeography and severity of end-Permian marine mass extinction. *Science* **362**, eaat1327 (2018).
268. Rey, K. *et al.* Oxygen isotopes suggest elevated thermometabolism within multiple Permo-Triassic therapsid clades. *eLife* **6**, (2017).
269. BioVisions. *BioVisions.* Harvard University. http://biovisions.mcb.harvard.edu/ (2019).
270. von Neumann, J. In *Probabilistic Logics and Synthesis of Reliable Organisms from Unreliable Components* (eds. C. Shannon and J. McCarthy) https://static.ias.edu/pitp/archive/2012files/Probabilistic_Logics.pdf (1952).
271. Atlan, H. Du bruit comme principe d'auto-organisation. *Communications* **18**, 21–36 (1972).
272. von Neumann, J. & Morgenstern, O. *Theory of Games and Economic Behavior.* (Princeton University Press, 1944).
273. *Stochastic Biomathematical Models.* Vol. 2058 (Springer Berlin Heidelberg, Berlin, Heidelberg, 2013).
274. Guptasarma, P. Does replication-induced transcription regulate synthesis of the myriad low copy number proteins of *Escherichia coli*? *BioEssays* **17**, 987–997 (1995).
275. FAO. *Wheat Yields from 1961 Onwards.* https://openknowledge.fao.org/bitstreams/97d0381f-8677-44f0-a5b3-020e50c6d8a2/download (2017).

276. Elowitz, M. B., Levine, A. J., Siggia, E. D. & Swain, P. S. Stochastic gene expression in a single cell. *Science* **297**, 1183–1186 (2002).
277. Heitzler, P. & Simpson, P. The choice of cell fate in the epidermis of Drosophila. *Cell* **64**, 1083–1092 (1991).
278. von Foerster, H. On self organizing systems and their environments. In *Self Organizing System*. (Yovitz & Cameron, Pergamon, 1960).
279. Zavala, E. & Marquez-Lago, T. T. Delays induce novel stochastic effects in negative feedback gene circuits. *Biophys. J.* **106**, 467–478 (2014).
280. Kaern, M., Elston, T. C., Blake, W. J. & Collins, J. J. Stochasticity in gene expression: From theories to phenotypes. *Nat. Rev. Genet.* **6**, 451–464 (2005).
281. Yu, J., Xiao, J., Ren, X., Lao, K. & Xie, X. S. Probing gene expression in live cells, one protein molecule at a time. *Science* **311**, 1600–1603 (2006).
282. Hänggi, P. Stochastic resonance in biology. How noise can enhance detection of weak signals and help improve biological information processing. *Chemphyschem Eur. J. Chem. Phys. Phys. Chem.* **3**, 285–290 (2002).
283. Tigges, M., Marquez-Lago, T. T., Stelling, J. & Fussenegger, M. A tunable synthetic mammalian oscillator. *Nature* **457**, 309–312 (2009).
284. Kupiec, J.-J. & Sonigo, P. *Ni Dieu Ni Gène: Pour Une Autre Théorie de l'hérédité*. (Seuil, Paris, 2000).
285. Kitano, H. Biological robustness. *Nat. Rev. Genet.* **5**, 826–837 (2004).
286. Mantovani, A. Redundancy and robustness versus division of labour and specialization in innate immunity. *Semin. Immunol.* **36**, 28–30 (2018).
287. Casanova, J.-L. & Abel, L. Human genetics of infectious diseases: Unique insights into immunological redundancy. *Semin. Immunol.* **36**, 1–12 (2018).
288. von Neumann, J. & Burk, A. W. *Theory of Self-Reproducing Automata*. https://cba.mit.edu/events/03.11.ASE/docs/VonNeumann.pdf (1966).
289. Todorov, E. & Jordan, M. I. Optimal feedback control as a theory of motor coordination. *Nat. Neurosci.* **5**, 1226–1235 (2002).
290. Wu, H. G., Miyamoto, Y. R., Gonzalez Castro, L. N., Ölveczky, B. P. & Smith, M. A. Temporal structure of motor variability is dynamically regulated and predicts motor learning ability. *Nat. Neurosci.* **17**, 312–321 (2014).
291. Singh, P., Jana, S., Ghosal, A. & Murthy, A. Exploration of joint redundancy but not task space variability facilitates supervised motor learning. *Proc. Natl. Acad. Sci. U. S. A.* **113**, 14414–14419 (2016).
292. Haff, P. K. Technology as a geological phenomenon: Implications for human well-being. *Geol. Soc. Lond. Spec. Publ.* **395**, 301–309 (2014).
293. Crowe, S. A. *et al.* Atmospheric oxygenation three billion years ago. *Nature* **501**, 535–538 (2013).
294. Luo, G. *et al.* Rapid oxygenation of Earth's atmosphere 2.33 billion years ago. *Sci. Adv.* **2**, e1600134 (2016).
295. Scheu, S. & Drossel, B. Sexual reproduction prevails in a world of structured resources in short supply. *Proc. Biol. Sci.* **274**, 1225–1231 (2007).
296. Uyttewaal, M. *et al.* Mechanical stress acts via katanin to amplify differences in growth rate between adjacent cells in Arabidopsis. *Cell* **149**, 439–451 (2012).
297. Aigouy, B. *et al.* Cell flow reorients the axis of planar polarity in the wing epithelium of Drosophila. *Cell* **142**, 773–786 (2010).
298. Heckel, E. *et al.* Oscillatory flow modulates mechanosensitive klf2a expression through trpv4 and trpp2 during heart valve development. *Curr. Biol. CB* **25**, 1354–1361 (2015).
299. Coen, E. & Rebocho, A. B. Resolving conflicts: Modeling genetic control of plant morphogenesis. *Dev. Cell* **38**, 579–583 (2016).
300. Hamant, O. *et al.* Developmental patterning by mechanical signals in Arabidopsis. *Science* **322**, 1650–1655 (2008).

301. Hervieux, N. *et al.* A mechanical feedback restricts sepal growth and shape in arabidopsis. *Curr. Biol.* **26**, 1019–1028 (2016).
302. Pan, Y., Heemskerk, I., Ibar, C., Shraiman, B. I. & Irvine, K. D. Differential growth triggers mechanical feedback that elevates Hippo signaling. *Proc. Natl. Acad. Sci. U. S. A.* **113**, E6974–E6983 (2016).
303. Hamant, O. & Moulia, B. How do plants read their own shapes? *New Phytol.* **212**, 333–337 (2016).
304. Wolpert, L. Arms and the man: The problem of symmetric growth. *PLoS Biol.* **8**, e1000477 (2010).
305. Hong, L. *et al.* Variable cell growth yields reproducible organ development through spatiotemporal averaging. *Dev. Cell* **38**, 15–32 (2016).
306. Arp, T. B. *et al.* Quieting a noisy antenna reproduces photosynthetic light-harvesting spectra. *Science* **368**, 1490–1495 (2020).
307. Ashby, W. R. Requisite variety and its implications for the control of complex systems. *Cybernetica* 83–99 (1958).
308. Broswimmer, F. J. *Ecocide: A Short History of Mass Extinction of Species.* (Pluto Press, London; Sterling, VA, 2002).
309. Novák, B. & Tyson, J. J. Design principles of biochemical oscillators. *Nat. Rev. Mol. Cell Biol.* **9**, 981–991 (2008).
310. Oates, A. C., Morelli, L. G. & Ares, S. Patterning embryos with oscillations: Structure, function and dynamics of the vertebrate segmentation clock. *Dev. Camb. Engl.* **139**, 625–639 (2012).
311. Forterre, Y. Slow, fast and furious: Understanding the physics of plant movements. *J. Exp. Bot.* **64**, 4745–4760 (2013).
312. Bastien, R., Bohr, T., Moulia, B. & Douady, S. Unifying model of shoot gravitropism reveals proprioception as a central feature of posture control in plants. *Proc. Natl. Acad. Sci. U. S. A.* **110**, 755–760 (2013).
313. Momiji, H. & Monk, N. A. M. Oscillatory Notch-pathway activity in a delay model of neuronal differentiation. *Phys. Rev. E Stat. Nonlin. Soft Matter Phys.* **80**, 021930 (2009).
314. Lee, T. I. *et al.* Transcriptional regulatory networks in *Saccharomyces cerevisiae*. *Science* **298**, 799–804 (2002).
315. Boyer, L. A. *et al.* Core transcriptional regulatory circuitry in human embryonic stem cells. *Cell* **122**, 947–956 (2005).
316. Vidal, E. A. *et al.* Nitrate-responsive miR393/AFB3 regulatory module controls root system architecture in Arabidopsis thaliana. *Proc. Natl. Acad. Sci.* **107**, 4477–4482 (2010).
317. Mangan, S., Itzkovitz, S., Zaslaver, A. & Alon, U. The incoherent feed-forward loop accelerates the response-time of the gal system of Escherichia coli. *J. Mol. Biol.* **356**, 1073–1081 (2006).
318. Goentoro, L., Shoval, O., Kirschner, M. W. & Alon, U. The incoherent feedforward loop can provide fold-change detection in gene regulation. *Mol. Cell* **36**, 894–899 (2009).
319. Hart, Y. *et al.* Paradoxical signaling by a secreted molecule leads to homeostasis of cell levels. *Cell* **158**, 1022–1032 (2014).
320. Milo, R. Network motifs: Simple building blocks of complex networks. *Science* **298**, 824–827 (2002).
321. Lander, E. S. *et al.* Initial sequencing and analysis of the human genome. *Nature* **409**, 860–921 (2001).
322. Pray, L. DNA replication and causes of mutation. *Nature education* 214 (2008).
323. Rajon, E. & Masel, J. Evolution of molecular error rates and the consequences for evolvability. *Proc. Natl. Acad. Sci.* **108**, 1082–1087 (2011).
324. Portman, J. R., Brouwer, G. M., Bollins, J., Savery, N. J. & Strick, T. R. Cotranscriptional R-loop formation by Mfd involves topological partitioning of DNA. *Proc. Natl. Acad. Sci.* **118**, e2019630118 (2021).

325. Boyle, C. A. & Cordero, J. F. Birth defects and disabilities: A public health issue for the 21st century. *Am. J. Public Health* **95**, 1884–1886 (2005).
326. Lucchese, V. Face aux risques, notre complexité nous rend vulnérables. *Usbek & Rica*. https://usbeketrica.com/article/face-aux-risques-notre-complexite-nous-rend-vulnerables (2019).
327. Tsubota, K. *et al.* Computer simulation of trabecular remodeling in human proximal femur using large-scale voxel FE models: Approach to understanding Wolff's law. *J. Biomech.* **42**, 1088–1094 (2009).
328. Dalous, J. *et al.* Reversal of cell polarity and actin-myosin cytoskeleton reorganization under mechanical and chemical stimulation. *Biophys. J.* **94**, 1063–1074 (2008).
329. Green, P. & King, A. A mechanism for the origin of specifically oriented textures in development with special reference to Nitella wall texture. *Aust. J. Biol. Sci.* 421–437 (1966).
330. Sapala, A. *et al.* Why plants make puzzle cells, and how their shape emerges. *eLife* **7**, (2018).
331. Moulia, B., Douady, S. & Hamant, O. Fluctuations shape plants through proprioception. *Science* **372**, eabc6868 (2021).
332. Allix, G. *Pour rendre aimable la ville dense, peut-être faut-il la vouloir moins parfaite.* (Le Monde, Paris, 2019).
333. Tocqueville, A. de & Raynaud, P. *De la démocratie en Amérique.* (Flammarion, Paris, 2010).
334. Agrobiobase. *Agrobiobase, the Showcase of Biobased Products.* http://www.agrobiobase.com/ (2019).
335. Kidd, P. S. *et al.* Developing sustainable agromining systems in agricultural ultramafic soils for nickel recovery. *Front. Environ. Sci.* **6** (2018).
336. Econick. *Econick.* https://econick.fr/.
337. Ghanem, M. A. *et al.* Growing phenotype-controlled phononic materials from plant cells scaffolds. *Appl. Mater. Today* **22**, 100934 (2021).
338. Hofte, H. *Biomass for the Future (BFF).* https://www.gisbiotechnologiesvertes.com/en/labeled-projects/biomass-for-the-future-bff-2 (2012).
339. Church, G. M., Gao, Y. & Kosuri, S. Next-generation digital information storage in DNA. *Science* **337**, 1628 (2012).
340. Extance, A. How DNA could store all the world's data. *Nature* **537**, 22–24 (2016).
341. Daddy, N. B. *et al.* Artemisia annua dried leaf tablets treated malaria resistant to ACT and i.v. artesunate: Case reports. *Phytomedicine* **32**, 37–40 (2017).
342. Hallé, F. *Éloge de la plante: pour une nouvelle biologie.* (Éd. Points, Paris, 2014).
343. Harari, Y. N. *Sapiens: A Brief History of Humankind.* (Harper, New York, 2015).
344. De Schutter, O. *Agroécologie et droit à l'alimentation.* Rapport présenté à la 16ème session du Conseil des droits de l'homme de l'ONU [A/HRC/16/49]. http://www.srfood.org/images/stories/pdf/officialreports/20110308_a-hrc-16-49_agroecology_en.pdf (2011).
345. Barot, S. *et al.* Designing mixtures of varieties for multifunctional agriculture with the help of ecology. A review. *Agron. Sustain. Dev.* **37**, 13 (2017).
346. Adu-Gyamfi, P., Mahmood, T. & Trethowan, R. Can wheat varietal mixtures buffer the impacts of water deficit? *Crop Pasture Sci.* **66**, 757 (2015).
347. Bouain, N. *et al.* Systems genomics approaches provide new insights into Arabidopsis thaliana root growth regulation under combinatorial mineral nutrient limitation. *PLOS Genet.* **15**, e1008392 (2019).
348. Schepman, T. *Tomates sans arrosage ni pesticide: cette méthode fascine les biologistes.* https://www.nouvelobs.com/rue89/rue89-planete/20160814.RUE8088/tomates-sans-arrosage-ni-pesticide-cette-methode-fascine-les-biologistes.html (2016).

349. Dupraz, C. *Agroforesterie: une nouvelle approche agricole*. https://www.fondationdefrance.org/fr/agroforesterie-une-nouvelle-approche-agricole (2018).
350. Kenyon, G. *BBC – Future – How Weeds Help Fight Climate Change*. http://www.bbc.com/future/story/20190507-weeds-a-surprising-way-to-fight-climate-change (2019).
351. Clément, G. *Jardins, Paysage et Génie Naturel*. (Collège de France, Fayard, Paris, 2012).
352. Le Lay, Y.-F. L'évaluation environnementale du bois en rivière par les gestionnaires des cours d'eau français. *Géocarrefour* **81**, 265–275 (2006).
353. Franz, C., Baser, K. & Windisch, W. Essential oils and aromatic plants in animal feeding – a European perspective. A review. *Flavour Fragr. J.* **25**, 327–340 (2010).
354. Foucault, C. Si Bayer rachète Monsanto, c'est aussi (et surtout) pour le digital. *usine-digitale.fr*. https://www.usine-digitale.fr/editorial/si-bayer-rachete-monsanto-c-est-aussi-et-surtout-pour-le-digital.N394597 (2016).
355. Griffon, M. *Nourrir la planète: pour une révolution doublement verte*. (O. Jacob, Paris, 2006).
356. Raworth, K. & Bury, L. *La théorie du donut: l'économie de demain en 7 principes*. (Plon, Paris, 2018).
357. Aristote & Voilquin, J. *Éthique de Nicomaque*. (Flammarion, Paris, 1998).
358. Hayek, F. A. von & Blumberg, G. *La route de la servitude*. (PUF, Paris, 2014).
359. Vosoughi, S., Roy, D. & Aral, S. The spread of true and false news online. *Science* **359**, 1146–1151 (2018).
360. Polanyi, K. *La grande transformation: aux origines politiques et économiques de notre temps*. (Gallimard, Paris, 2017).
361. Choler, P., Michalet, R. & Callaway, R. M. Facilitation and competition on gradients in alpine plant communities. *Ecology* **82**, 3295–3308 (2001).
362. Hoek, T. A. *et al.* Resource availability modulates the cooperative and competitive nature of a microbial cross-feeding mutualism. *PLOS Biol.* **14**, e1002540 (2016).
363. Stiglitz, J. Neoliberalism must be pronounced dead and buried. Where next? | Joseph Stiglitz. *The Guardian* (2019).
364. CIA. *Le monde en 2035 vu par la CIA et le Conseil national du renseignement: le paradoxe du progrès: document*. https://ia803409.us.archive.org/29/items/le-monde-en-2035-vu-par-la-cia_202105/Le%20monde%20en%202035%20vu%20par%20la%20CIA.pdf (2018).
365. Platt, J. Social traps. *Am. Psychol.* **28**, 641–651 (1973).
366. Darwin, C. *La filiation de l'homme et la sélection liée au sexe*. (H. Champion, Paris, 1871).
367. Hobbes, T. *Léviathan*. (1651).
368. Barrière, O. *et al. Coviabilité Des Systèmes Sociaux et Écologiques*. (Ed. Matériologiques, Paris, 2019).
369. Council of the City of Toledo. *Lake Erie Bill of Rights*. https://www.utoledo.edu/law/academics/ligl/pdf/2019/Lake-Erie-Bill-of-Rights-GLWC-2019.pdf (2019).
370. Lordon, F. Pleurnicher le Vivant. *Le Monde diplomatique* (2021).
371. Hardin, G. The Tragedy of the Commons. *Science* **162**, 1243–1248 (1968).
372. Ostrom, E. *Governing the Commons: The Evolution of Institutions for Collective Action*. (Cambridge University Press, Cambridge; New York, 1990).
373. Mitra, S. *The School in the Cloud: The Emerging Future of Learning*. (Corwin, Thousand Oaks, CA, 2020).
374. Querrien, A. *L'école Mutuelle: Une Pédagogie Trop Efficace*. (Empêcheurs de penser en rond, Paris, 2005).
375. Bregman, R. *Rutger Bregman: La pauvreté n'est pas un manque de caractère, c'est un manque d'argent | TED Talk*. https://www.ted.com/talks/rutger_bregman_poverty_isn_t_a_lack_of_character_it_s_a_lack_of_cash?language=fr (2017).

376. Mullainathan, S. & Shafir, E. *Scarcity: The New Science of Having Less and How It Defines Our Lives*. (Picador/Henry Holt and Co, New York, 2014).
377. Tierney, J. Do you suffer from decision fatigue? *The New York Times* (2011).
378. Forget, E. L. The town with no poverty: The health effects of a Canadian guaranteed annual income field experiment. *Can. Public Policy* **37**, 283–305 (2011).
379. Jauch, H. NAMIBIE. Les miracles du revenu minimum garanti. *Courrier international*. https://www.courrierinternational.com/article/2010/04/29/les-miracles-du-revenu-minimum-garanti (2010).
380. Choisne, C. *Discours Remise des Diplômes 2018 Centrale Nantes*. https://www.youtube.com/watch?v=3LvTgiWSAAE&vl=fr (2018).
381. Dalmais, M. Pourquoi je refuse mon diplôme d'Ingénieur. Conférence gesticulée de Mathieu Dalmais. *Ingénieurs sans frontières*. https://www.isf-france.org/videos/pourquoi-je-re-fuse-mon-diplome-dingenieur-conference-gesticulee-de-mathieu-dalmais (2016).
382. Izoard, C. *Merci de changer de métier: lettres aux humains qui robotisent le monde*. (Les Ami-e-s de Clark Kent, Montreuil, 2020).
383. Bonnet, E., Landivar, D. & Monnin, A. Crise climatique: " Nous devons apprendre à désinnover ". *Le Monde.fr* (2021).
384. Graner, F. Devons-nous arrêter la recherche? *Pièces et Main d'Oeuvre*. https://www.piecesetmaindoeuvre.com/spip.php?page=resume&id_article=1573 (2021).
385. MFRB. Mouvement Français pour un Revenu de Base – Accueil. *Mouvement Français pour un Revenu de Base*. https://www.revenudebase.info/ (2019).
386. Friot, B. & Zech, P. *Emanciper le travail: entretiens avec Patrick Zech*. (La Dispute, Paris, 2014).
387. Bayon, D. *Le commerce véridique et social de Michel-Marie Derrion: Lyon, 1835–1838: petites visites chez les utopies coopératives nos grand-parents*. (Atelier de Création Libertaire, Lyon, 2002).
388. Terre de Liens. *Terre de Liens – Et si vous faisiez pousser des fermes?* https://terredeliens.org/./ (2003).
389. Bregman, R. *Rutger Bregman Tells Davos to Talk about Tax: 'This Is Not Rocket Science'*. https://www.theguardian.com/business/video/2019/jan/30/this-is-not-rocket-science-rutger-bregman-tells-davos-to-talk-about-tax-video (2019).
390. Piketty, T. *Capital et Idéologie*. (Seuil, Paris, 2019).
391. Orbell, J. M., Van de Kragt, A. J. & Dawes, R. M. Explaining discussion-induced cooperation. *J. Pers. Soc. Psychol.* **54**, 811–819 (1988).
392. Van Vugt, M. & De Cremer, D. Leadership in social dilemmas: The effects of group identification on collective actions to provide public goods. *J. Pers. Soc. Psychol.* **76**, 587–599 (1999).
393. CCC. Site officiel de la Convention Citoyenne pour le Climat. *Convention Citoyenne pour le Climat*. https://www.conventioncitoyennepourleclimat.fr/ (2019).
394. King, M. W. *How Brain Biases Prevent Climate Action*. https://www.bbc.com/future/article/20190304-human-evolution-means-we-can-tackle-climate-change (2019).
395. Bourg, D. *Pour Une 6e République Écologique*. (Odile Jacob, Paris, 2011).
396. Stengers, I. *Réactiver Le Sens Commun: Lecture de Whitehead En Temps de Débâcle*. (Éditions La Découverte, Paris, 2020).
397. Piketty, T. Pour une économie circulaire. *Le Monde.fr* (2019).
398. Ostrom, E. A general framework for analyzing sustainability of social-ecological systems. *Science* **325**, 419–422 (2009).
399. Rifkin, J. *The Third Industrial Revolution: How Lateral Power Is Transforming Energy, the Economy, and the World*. (Palgrave Macmillan, Basingstoke, 2013).
400. ZAD. Prise de terre(s). *lundimatin*. https://lundi.am/ZAD (2019).
401. Valo, M. Pour protéger la qualité de son eau potable, Paris tente de convertir les agriculteurs au bio. *Le Monde* (2019).

402. Pointereau, R. & Bourquin, M. *Rapport d'information sur la revitalisation des centres-villes et des centres-bourgs.* https://www.senat.fr/rap/r16-676/r16-676.html (2017).
403. Liss, D. Comment Mulhouse est devenu l'un des centres-villes les plus dynamiques de France. *Rue89 Strasbourg.* https://www.rue89strasbourg.com/mulhouse-centres-ville-dynamique-136234 (2018).
404. Hipp, A.-K. Chute du mur. 1989–2019: Berlin, capitale du changement perpétuel. *Courrier international/Der Tagesspiegel.* https://www.courrierinternational.com/article/chute-du-mur-1989-2019-berlin-capitale-du-changement-perpetuel (2019).
405. Lucchese, V. Les communes sont une très bonne échelle pour la transition écologique. *Usbek & Rica.* https://usbeketrica.com/article/les-communes-sont-une-tres-bonne-echelle-pour-la-transition-ecologique (2019).
406. Hers, F. *Les Nouveaux commanditaires.* http://www.nouveauxcommanditaires.eu/ (2019).
407. Frumkin, H., Hess, J. & Vindigni, S. Energy and public health: The challenge of peak petroleum. *Public Health Rep.* **124**, 5–19 (2009).
408. Dalglish, S. L., Poulsen, M. N. & Winch, P. J. Localization of health systems in low- and middle-income countries in response to long-term increases in energy prices. *Glob. Health* **9**, 56 (2013).
409. *BedZED (Beddington Zero Energy Development).* The BedZED Story: The UK's First Large-Scale, Mixed-Use Eco-Village. (2017).
410. Ginsberg, J. *et al.* Detecting influenza epidemics using search engine query data. *Nature* **457**, 1012–1014 (2009).
411. Lazer, D., Kennedy, R., King, G. & Vespignani, A. The parable of Google flu: Traps in big data analysis. *Science* **343**, 1203–1205 (2014).
412. Caliskan, A., Bryson, J. J. & Narayanan, A. Semantics derived automatically from language corpora contain human-like biases. *Science* **356**, 183–186 (2017).
413. Kosinski, M., Stillwell, D. & Graepel, T. Private traits and attributes are predictable from digital records of human behavior. *Proc. Natl. Acad. Sci.* **110**, 5802–5805 (2013).
414. Flandrin, A. *Comment Trump a manipulé l'Amérique.* (Le Monde, Paris, 2018).
415. Floret, Pr. *Plan d'elimination de la rougeole et de la rubeole congenitale en France.* https://sante.gouv.fr/IMG/pdf/plan_elimination_rougeole.pdf (2005).
416. Wolff, F. *Plaidoyer Pour l'universel: Fonder l'humanisme.* (Fayard, Paris, 2019).
417. Smithers, R. Scariest thing about Halloween is plastic waste, say charities. *The Guardian* (2019).
418. Petit, V. Transition écologique et numérique. Vers des territoires communs? *Rev. DEconomie Reg. Urbaine* **Décembre**, 797–818 (2017).
419. Latouche, S. *Farewell to Growth.* (Polity, Cambridge; Malden, MA, 2009).
420. Mendoza, C. & Pollet, J.-F. Le buen vivir, un petit laboratoire social importé du Sud. *CETRI: Centre tricontinental.* https://www.cetri.be/Le-buen-vivir-un-petit-laboratoire (2021).
421. Gandhi, M. *"Hind Swaraj" and Other Writings* (ed. Parel, A. J.) (Cambridge University Press, Cambridge; New York, 1997).
422. Muzaffar, M. Disenchanted young Chinese are 'lying flat' to protest work and hustle culture. *The Independent.* https://www.independent.co.uk/asia/china/china-tang-ping-trend-work-culture-b1862444.html (2021).
423. UNDP. *Human Development Report 1990.* https://hdr.undp.org/content/human-development-report-1990 (1990).
424. Hickel, J. The sustainable development index: Measuring the ecological efficiency of human development in the Anthropocene. *Ecol. Econ.* **167**, 106331 (2020).
425. Bonnet, E., Landivar, D. & Monnin, A. *Héritage et fermeture: une écologie du démantèlement.* (Divergences, Paris, 2021).
426. D'Allessandro, S., Dittmer, K., Distefano, T. & Cieplinski, A. *EUROGREEN Model of Job Creation in a Post-growth Economy.* https://people.unipi.it/simone_dalessandro/wp-content/uploads/sites/78/2018/10/EUROGREEN_Project.pdf (2018).

427. Martínez Franzoni, J. & Sánchez-Ancochea, D. *The Quest for Universal Social Policy in the South: Actors, Ideas and Architectures*. (Cambridge University Press, London, 2016).
428. Frey, P. *The Ecological Limits of Work: On Carbon Emissions, Carbon Budgets and Working Time*. https://ucc.socioeco.org/bdf_fiche-document-6687_en.html (2019).
429. Orton, J. D. & Weick, K. E. Loosely coupled systems: A reconceptualization. *Acad. Manage. Rev.* **15**, 203–223 (1990).
430. Weick, K. Educational organizations as loosely coupled systems. *Adm. Sci. Q.* 1–19 (1976).
431. Goldratt, E. M., Cox, J. & Miremont, J.-C. *Le but: un processus de progrès permanent*. (Afnor éditions, La Plaine Saint-Denis, 2017).
432. Laloux, F. *Reinventing Organizations: A Guide to Creating Organizations Inspired by the Next Stage of Human Consciousness*. (Nelson Parker, Brussels, 2014).
433. Huxley, A. *Le meilleur des mondes*. (Pocket, Paris, 1932).
434. Boltanski, L. & Chiapello, È. *Le nouvel esprit du capitalisme*. (Gallimard, Paris, 2011).
435. Chapoutot, J. *Libres d'obéir: Le Management, Du Nazisme à Aujourd'hui*. (Gallimard, Paris, 2020).
436. Labarthe, A. S. *Georges Bataille à perte de vue*. (Les Films du Bief, Paris, 1997).
437. Rosa, H. & Mannoni, O. *Rendre le monde indisponible*. (la Découverte, Paris, 2020).
438. Bouchain, P. L'harmonie, c'est ce qui fait la beauté de l'architecture. *France Culture*. https://www.franceculture.fr/emissions/linvite-des-matins/imaginer-la-ville-de-demain-patrick-bouchain-est-linvite-des-matins (2020).
439. Hamant, O. & Vallat, D. Plus une ville est optimisée, moins elle est résiliente, car il n'existe pas de ressources cachées pour s'adapter et rebondir. *Le Monde* (2020).
440. Podalydes, D. Denis Podalydès: 'Il faut consentir à ses moments désespérés pour réussir à faire sortir l'énergie'. *France Culture*. https://www.franceculture.fr/emissions/les-masterclasses/denis-podalydes-tout-acteur-doit-savoir-rater-de-facon-a-laver-le-corps-et-lesprit-de-tout-ce-quil (2020).
441. Hessel, S. *Indignez-vous!* (Indigène éd, Montpellier, 2010).
442. Mertens, J. En France, Netflix représente près de 25% du trafic sur le web. *Geeko*. https://geeko.lesoir.be/2019/07/01/en-france-netflix-represente-pres-de-25-du-trafic-sur-le-web/ (2019).
443. Nau, J.-Y. "Parker", "parkerisé" et "parkérisation". *Slate.fr*. http://www.slate.fr/story/8659/les-mots-du-vin (2009).
444. Labov, W. Unendangered dialect, endangered people: The case of African American vernacular English. *Transform. Anthropol.* **18**, 15–27 (2010).
445. Kohn, E. & Descola, P. *Comment pensent les forêts: vers une anthropologie au-delà de l'humain*. (Zones sensibles, Bruxelles, 2017).
446. Wohlleben, P. *The Hidden Life of Trees: What They Feel, How They Communicate – Discoveries from a Secret World*. (Greystone, Vancouver, Canada, 2016).
447. Grumbach, S. & Hamant, O. How humans may co-exist with Earth? The case for suboptimal systems. *Anthropocene* **30**, 100245 (2020).
448. Raynaud, R. *Fantasy et écologie: les mondes sous-optimaux de Robin Hobb et Pierre Bottero*. (Master thesis, Lyon, 2020).
449. Petrini, C. *Slow Food Nation: Why Our Food Should Be Good, Clean, and Fair*. (Rizzoli Ex Libris, New York, NY, 2013).
450. Fournier, C. Slow TV, slow food, slow cities . . . Cinq façons de ralentir le rythme. *Franceinfo*. https://www.francetvinfo.fr/culture/slow-tv-slow-food-slow-cities-cinq-facons-de-ralentir-le-rythme_469634.html (2013).
451. Pyta, A. *et al. The Longing*. (Studio Seufz, Heidelberg, 2019).
452. Moshfegh, O., Baude, C. & Moshfegh, O. *Mon Année de Repos et de Détente: Roman*. (Fayard, Paris, 2019).
453. Altieri, M. A. & Pimbert, M. *L'agroécologie: bases scientifiques d'une agriculture alternative*. (C. Corlet, Condé-sur-Noireau, 2013).

454. Smil, V. *Your Phone Costs Energy – Even before You Turn It on.* https://trends.directindustry.com/project-112007.html (2016).
455. Mazzucato, M. *The Entrepreneurial State: Debunking Public vs. Private Sector Myths.* (Anthem Press, London; New York, 2014).
456. Morin, E. *Science Avec Conscience.* (Éd. du Seuil, Paris, 1990).
457. Vallat, D. Apprivoiser les cygnes noirs: enseignements de la crise du coronavirus. *The Conversation.* http://theconversation.com/apprivoiser-les-cygnes-noirs-enseignements-de-la-crise-du-coronavirus-135481 (2020).
458. Levinas, E. *Le temps et l'autre.* (PUF, Paris, 1947).
459. Mintzberg, H. *Le manager au quotidien: les dix rôles du cadre.* (Organisation, Paris, 2016).
460. Queval, I. *S'accomplir Ou Se Dépasser: Essai Sur Le Sport Contemporain.* (Gallimard, Paris, 2004).
461. Tversky, A. & Kahneman, D. Judgment under Uncertainty: Heuristics and Biases. *Science* **185**, 1124–1131 (1974).
462. Bernoulli, D. *La Théorie de l'utilité.* (1738).
463. Horvilleur, D., Gorog, F., Faucher, L. & Habib, S. *Le rabbin et le psychanalyste: l'exigence d'interprétation.* (Hermann, Paris, 2020).
464. Al-Farabi, A. N. M. *Opinions des habitants de la cité vertueuse.* (Albouraq, Paris, 2011).
465. Obadia, C. *L'homme Inachevé.* (Maïa, Paris, 2021).

Index

D

E

F

For Product Safety Concerns and Information please contact our EU representative GPSR@taylorandfrancis.com Taylor & Francis Verlag GmbH, Kaufingerstraße 24, 80331 München, Germany